Reliability Calculations with the Stochastic Finite Element

Authored by

Wenhui Mo

School of Mechanical Engineering
Hubei University of Automotive Technology
China

Reliability Calculations with the Stochastic Finite Element

Author: Wenhui Mo

ISBN (Online): 978-981-14-8553-4

ISBN (Print): 978-981-14-8551-0

ISBN (Paperback): 978-981-14-8552-7

need for a court order if at any point you breach any terms of this License Agreement. In no event will any delay or failure by Bentham Science Publishers in enforcing your compliance with this License Agreement constitute a waiver of any of its rights.

3. You acknowledge that you have read this License Agreement, and agree to be bound by its terms and conditions. To the extent that any other terms and conditions presented on any website of Bentham Science Publishers conflict with, or are inconsistent with, the terms and conditions set out in this License Agreement, you acknowledge that the terms and conditions set out in this License Agreement shall prevail.

Bentham Science Publishers Pte. Ltd.
80 Robinson Road #02-00
Singapore 068898
Singapore
Email: subscriptions@benthamscience.net

CONTENTS

PREFACE

There are two kinds of uncertainties, fussiness and randomness in engineering problems. Several researchers in China and abroad pay attention to the influence of random factors on the structure. In machinery, dam, construction, earthquake and other fields, random factors do have a great impact on the structure. The spatial variability of structural material properties is studied as a random process by many scholars. With the deepening of human understanding, it is not practical to ignore the design of randomness.

In the first chapter, the fuzzy reliability of a single component maintenance system and the repairable series system are studied. Two fuzzy methods for reliability allocation are proposed. The second chapter discusses the reliability of the rigid rotor balance. Based on the sensitivity analysis, a Monte Carlo simulation for the reliability calculation of gears is proposed. Based on the sensitivity analysis, an optimization method for reliability calculation is proposed. The reliability calculation of spring is studied by using the HL-RF method. In the third chapter, optimization design-based HL-RF and IS for the gearbox are proposed. A multi-objective reliability-based fuzzy optimization design for gear box is proposed.

The fourth chapter proposes an improved method of perturbation stochastic finite element to save computational time. In the fifth chapter, differential equations are transformed into linear equations by the Wilson θ method. Linear equations are solved by the Successive Over Relaxation method. Anew method of calculating dynamic reliability using the Neumann stochastic finite element is proposed. The sixth chapter discusses the design model of the gearbox established by using the stochastic finite element method. A new method of stochastic finite element for vibration is also proposed. In the last chapter, four stochastic finite element methods are proposed to calculate nonlinear vibration.

Wenhui Mo
School of Mechanical Engineering
Hubei University of Automotive Technology
China

CHAPTER 1

Fuzzy Reliability

Abstract: Considering the influence of fuzzy factors, the fuzzy reliability of single component maintenance system and the repairable series system is studied.

Two fuzzy methods for reliability allocation are proposed: one uses the second-order fuzzy comprehensive evaluation method, and the other one uses the fuzzy optimization method.

Keywords: Fuzzy reliability, Fuzzy optimization method, Repairable series system, Reliability allocation, Second-order fuzzy comprehensive evaluation method, Single component maintenance system.

INTRODUCTION

There is no absolute clear boundary between normal and abnormal operation (failure) of the system, but it is often a form of transition through an intermediary - work with failure, so it is a fuzzy concept. Zadeh LA, an American fuzzy mathematician, uses the degree of membership to describe the intermediary transition of differences, which is a description of fuzziness in precise mathematical language.

The fuzzy reliability analysis in the posits reliability theory is defined precisely and a general approach by a system of functional equations is proposed [1]. A fuzzy fault-tree based reliability analysis of an optimally planned transmission system is presented [2]. An attempt has been made to present a new approach for the stability analysis of slopes incorporating fuzzy uncertainty [3]. Fuzzy numbers are used to define an equivalence class of probability distributions compatible with available data and corresponding upper and lower cumulative density functions [4]. The most relevant parameters are identified by means of different sensitivity analysis techniques. Then, fuzzy models are devised which efficiently do the required mapping between the system outputs and the identified relevant inputs [5]. A new fuzzy multi-objective optimization method is introduced, and it is used for the optimization decision making of the series and complex system reliability with two goals [6]. An approach to fuzzy rule base design using a tabu search algorithm (TSA) for nonlinear system modeling is presented [7]. A new modelling approach for determining the reliability and availability of a production system is proposed by considering all the components of the system and their hierarchy in the system structure [8]. A new algorithm has been introduced to build the membership function and non-membership function of the fuzzy reliability of a system having components following different types of intuitionistic fuzzy failure rates [9]. A fuzzy-based reliability approach is presented to check the basic events of system fault trees, the failure precise probability distributions of which is not available [10]. Some recent results on the application of the fuzzy Bayes methodology for the analysis of imprecise reliability data are proposed [11]. New means for predicting time to failure of the components, using a calibration regression method for measuring the error prediction in the extrapolation process are proposed [12]. A road-map has been provided to assess the reliability indices of repairable systems with uncertain limits [13]. This paper introduces evidence variables and fuzzy variables simultaneously to describe the uncertain epistemic parameters and a novel dual-stage reliability analysis framework [14].

This chapter discusses the fuzzy reliability calculation of a single part repair system and series repair system. The second-order fuzzy comprehensive evaluation method and fuzzy optimization method for reliability allocation are proposed. The fuzzy method of reliability allocation of series system, parallel system and hybrid system is discussed in detail.

FUZZY RELIABILITY OF SINGLE COMPONENT MAINTENANCE SYSTEM

A single component maintenance system is the simplest maintenance system. The system is composed of one component. It starts to work from $t = 0$. In case of failure during a period of work, the repairman will immediately carry out maintenance, and then carry out normal work after repair.

Transition Fuzzy Probability

The life distribution of a single component is an exponential distribution with parameter λ

$$F(t) = P(T \leq t) = 1 - e^{-\lambda t} \quad t \geq 0 \quad \lambda > 0 \tag{1}$$

The repair time distribution of a single component is an exponential distribution with parameter μ

$$M(t) = P(\tau \leq t) = 1 - e^{-\mu t} \quad t \geq 0 \quad \mu > 0 \tag{2}$$

The state "0" indicates that the system is normal, the state "0 1" indicates that the system is working with failure, and the state "1" indicates that the system is in a fault state. Then, x (T) is used to show the state of the at time t. Let

$$x(t) = \begin{cases} 0 & \text{system works at time t} \\ 0 \quad 1 & \text{system works with fault at time t} \\ 1 & \text{system is in fault at time t} \end{cases}$$

$\{x$ (T)$, t \geq 0\}$ process is a fuzzy Markov process.

The membership degree of "0" state transferred to "0" state is ω_{00}

The membership degree of "0" state transferred to "1" state is ω_{01}

The membership degree of "1" state transferred to "0" state is ω_{10}

The membership degree of "1" state transferred to "1" state is ω_{11}

If the system is in working state at time t, and the system is in working state after time Δt, the fuzzy probability of the system from "0" state to "0" state in time interval Δt is

$$P_{00}(\Delta t) = P\{x(t + \Delta t) = 0 \mid x(t) = 0\} = \omega_{00} e^{-\lambda \Delta t}$$

$$= \omega_{00}(1 - \lambda \Delta t) + 0(\Delta t) \tag{3}$$

When the system is in working state at time t, and after time Δ t, the system fails, then the transition fuzzy probability of system from "0" state to "1" state in Δ t is

$$P_{01}(\Delta t) = P\{x(t + \Delta t) = 1 \mid x(t) = 0\} = \omega_{01}(1 - e^{-\lambda \Delta t}) = \lambda \omega_{01} \Delta t + 0(\Delta t) \tag{4}$$

When the system is in fault state at time t and is repaired within time interval Δt, after time Δt, the system transfers from the state "1" to "0", then the transfer fuzzy probability of the system is

$$P_{10}(\Delta t) = P\{x(t+\Delta t)=0 \,|\, x(t)=1\}$$

$$= \omega_{10}(1-e^{-\mu\Delta t}) = \mu\omega_{10}\Delta t + 0(\Delta t) \tag{5}$$

When the system is in fault state at time t, the system is still in fault state without repair in Δt, that is, the system transfers from "1" state to "1" state in Δt, then the transfer fuzzy probability of the system is

$$P_{11}(\Delta t) = P\{x(t+\Delta t)=1 \,|\, x(t)=1\} = \omega_{11}e^{\mu\Delta t}$$

$$= \omega_{11}(1-\mu\Delta t) + 0(\Delta t) \tag{6}$$

If the higher-order infinitesimal is ignored, the transition fuzzy probability of a state in time Δt is

$$P = \begin{pmatrix} \omega_{00}(1-\lambda\Delta t) & \lambda\omega_{01}\Delta t \\ \omega_{10}\mu\Delta t & \omega_{11}(1-\mu\Delta t) \end{pmatrix} \tag{7}$$

Fuzzy Validity of the System

The fuzzy validity $A(t)$ is to find the fuzzy probability of the system at time t in the "0" state

$$A(t) = P_0(t) = P\{x(t)=0\} \tag{8}$$

For the sake of solving $P_0(t)$, consider $P_1(t)$

$$P_1(t) = P\{x(t)=1\} \tag{9}$$

Then

$$P_0(t+\Delta t) = P_0(t)P_{00}(\Delta t) + P_1(t)P_{10}(\Delta t) \tag{10}$$

Substituting Eq. 3 and Eq. 5, we obtain

$$P_0(t+\Delta t) = \omega_{00}(1-\lambda\Delta t)P_0(t) + \omega_{10}\mu\Delta t P_1(t) + 0(\Delta t) \tag{11}$$

We can also get

$$P_1(t+\Delta t) = \lambda\omega_{01}\lambda\Delta t P_0(t) + \omega_{11}(1-\mu\Delta t)P_1(t) + 0(\Delta t) \tag{12}$$

When $\Delta t \to 0$, $\omega_{00} \to 1$, $\omega_{11} \to 1$, the differential equation of the system state and the initial conditions satisfied are obtained

$$\begin{cases} P_0'\ (t) = -\lambda\omega_{00}P_0(t) + \mu\omega_{10}P_1(t) \\ P_1'\ (t) = \lambda\omega_{01}P_0(t) + \mu\omega_{11}P_1(t) \\ P_0\ (0) = 1,\ \ P_1\ (0) = 1 \end{cases} \tag{13}$$

The system of equations (13) is a system of linear differential equations with constant coefficients, and the corresponding second-order linear differential equations with constant coefficients are

$$P_0''\ (t) + (\omega_{00}\lambda + \mu\omega_{11})P_0'(t) + \lambda\mu(\omega_{00}\omega_{11} - \omega_{01}\omega_{10})P_0(t) = 0$$

To solve it, we get

$$P_0(t) = C_1 e^{k_1 t} + C_2 e^{k_2 t} \tag{14}$$

where

$$k_1 = \left[-(\omega_{00}\lambda + \omega_{11}\mu) - \sqrt{(\omega_{00}\lambda - \omega_{11}\mu)^2 + 4\mu\lambda\omega_{01}\omega_{10}} \right] / 2 \tag{15}$$

$$k_2 = \left[-(\omega_{00}\lambda + \omega_{11}\mu) + \sqrt{(\omega_{00}\lambda - \omega_{11}\mu)^2 + 4\mu\lambda\omega_{01}\omega_{10}} \right]/2 \tag{16}$$

By substituting Eq. (14) in the first equation of equation system (13), we get

$$p_1(t) = \left[(c_1 k_1 + c_1\lambda\omega_{00}) / \omega_{10}\mu \right] e^{k_1 t} + \left[(c_2 k_2 + c_2\lambda\omega_{00}) / \omega_{10}\mu \right] e^{k_2 t} \tag{17}$$

By substituting the initial conditions in Eq. (14) and Eq. (17), we can get

$$C_1 = (-k_2 - \lambda\omega_{00}) / (k_1 - k_2) \tag{18}$$

$$C_2 = (k_2 + \lambda\omega_{00}) / (k_1 - k_2) \tag{19}$$

The fuzzy validity of the system is

$$A(t) = \left[(-k_2 - \lambda\omega_{00}) / (k_1 - k_2) \right] e^{k_1 t} + \left[(k_1 + \lambda\omega_{00}) / (k_1 - k_2) \right] e^{k_2 t} \tag{20}$$

Fuzzy Reliability of the System

Make the fault state "1" of the system as an absorption state, that is $\mu = 0$ (when the system fails, it will not be repaired). In this way, a new fuzzy Markov process $\{Y(t), t \geq 0\}$ is formed

$$R(t) = P_0(t) = P\{Y(t) = 0\} \tag{21}$$

Similar to the previous derivation, the equations are obtained

$$P_0(t+\Delta t) = P_0(t)P_{00}(\Delta t) = \omega_{00}(1 - \lambda\Delta t)P_0(t) + O(\Delta t) \tag{22}$$

$$P_1(t+\Delta t) = P_0(t)P_{01}(\Delta t) + P_1(t)P_{11}(\Delta t) = \omega_{01}\lambda\Delta t P_0(t) + \omega_{11}P_1(t) + O(\Delta t) \tag{23}$$

When $\Delta t \to 0$, the differential equations and the initial conditions are

$$\begin{cases} P_0'(t) = -\omega_{00}\lambda P_0(t) \\ P_1'(t) = \omega_{01}\lambda P_0(t) \\ P_0(0) = 1, P_1(0) = 0 \end{cases} \qquad (24)$$

Solutions of differential equations (24) satisfying initial conditions are

$$P_0(t) = e^{-\omega_{00}\lambda t}, \qquad P_1(t) = -(\omega_{01}/\omega_{00})e^{-\omega_{00}\lambda t} + (\omega_{01}/\omega_{00})$$

Therefore, the fuzzy reliability of the system is

$$R(t) = e^{-\omega_{00}\lambda t} \qquad (25)$$

Example

The life distribution of a single component is an exponential distribution. $\omega_{00} = 0.98$, $\omega_{01} = 0.9$, $\omega_{10} = 0.8$, $\omega_{11} = 0.98$, $\lambda = 0.0051/\text{h}$. If the system is not repairable, try to calculate the fuzzy reliability of the system for 50 hours. If the system is repairable, the repair rate is $\mu = 0.067/\text{h}$, then try to calculate the fuzzy validity of 50 hours' work.

According to Eq. (25), $R(50) = e^{-0.98 \times 0.0051 \times 50} = 0.7789$

According to Eq. (20), where, $k_1 = -0.069474, k_2 = -0.001184$, we can obtain

$A(50) = 0.892$

Fuzzy Reliability of Repairable Series System

The repairable system is affected by many fuzzy factors, such as design level, manufacturing level, material quality, and repair level. In this paper, the fuzzy validity and fuzzy reliability of the repairable series system are studied.

State and Membership Degree

$X(t)$ is used to represent the state of the system at time t.

$$x(t) = \begin{cases} 0 & \text{system works at time t} \\ 0 \quad 1 & \text{system works with fault at time t} \\ 1 & \text{system is in fault at time t} \end{cases}$$

{x (T), t ≥ 0} process is a fuzzy Markov process.

The membership degree of "0" state transferred to "0" state is ω_{00}. The membership degree of "0" state transferred to "1" state is ω_{01}. The membership degree of "1" state transferred to "0" state is ω_{10}. The membership degree of "1" state transferred to "1" state is ω_{11}.

Repairable Series System with the Same n Components

The repairable series system with the same n components consists of a repairman and the same n units in series. The distribution function of the life T_i of each unit is $1 - e^{-\lambda t}$, and the distribution function of the repair time of each unit after failure is $1 - e^{-\mu t}$, $i = 1, 2, \cdots, n$.

The transition fuzzy probability is as follows:

$$P_{00}(\Delta t) = P\{X(t+\Delta t) = 0 \,|\, X(t) = 0\}$$

$$= \omega_{00}(e^{-\lambda \Delta t})^n = \omega_{00}(1 - n\lambda\Delta t) + 0(\Delta t) \tag{26}$$

$$P_{01}(\Delta t) = P\{X(t+\Delta t) = 1 \,|\, X(t) = 0\}$$

$$= \omega_{01}\left[1 - (e^{-\lambda\Delta t})^n\right] = n\lambda\omega_{01}\Delta t + 0(\Delta t) \tag{27}$$

$$P_{10}(\Delta t) = P\{X(t+\Delta t) = 0 \,|\, X(t) = 1\}$$

$$= P\{\text{Faulty component repair in } (t,t+\Delta t) \mid \text{There is a component failure at time t}\}$$

$$= \mu\omega_{10}\Delta t + 0(\Delta t) \tag{28}$$

$$P_{11}(\Delta t) = P\{X(t+\Delta t)=1 \mid X(t)=1\} = P\{\text{The faulty parts have not been repaired in}$$
$$(t,t+\Delta t) \mid \text{There is a component failure at time t}\}$$

$$= \omega_{11}(1-\mu\Delta t) + 0(\Delta t) \tag{29}$$

$$P_i(t) = P\{X(t)=i\} \quad i = 0,1 \tag{30}$$

Hence,

$$P_0(t+\Delta t) = P_0(t)P_{00}(\Delta t) + P_1(t)P_{10}(\Delta t) \tag{31}$$

Substituting Eq. 26 and Eq. 28in the above equation, we can get

$$P_0(t+\Delta t) = \omega_{00}(1-n\lambda\Delta t)P_0(t) + \mu\omega_{10}P_1(t)\Delta t + 0(\Delta t) \tag{32}$$

We can also get

$$P_1(t+\Delta t) = n\lambda\omega_{01}p_0(t)\Delta t + \omega_{11}(1-\mu\Delta t)P_1(t) + 0(\Delta t) \tag{33}$$

When $\Delta t \to 0$, $\omega_{00} \to 1$, $\omega_{11} \to 1$, the differential equation of the system state and the initial conditions satisfied are obtained

$$\begin{cases} P'_0(t) = -n\lambda\omega_{00}P_0(t) + \mu\omega_{10}P_1(t) \\ P'_1(t) = n\lambda\omega_{01}P_0(t) - \mu\omega_{11}P_1(t) \\ P_0(0) = 1, P_1(0) = 0 \end{cases} \tag{34}$$

The system of equations Eq. 34 is a system of linear differential equations with constant coefficients, and the corresponding second-order linear differential equations with constant coefficients are as follows:

$$P_0''(t) + (n\lambda\omega_{00} + \mu\omega_{11})P_0'(t) + n\lambda\mu(\omega_{00}\omega_{11} - \omega_{01}\omega_{10})P_0(t) = 0 \tag{35}$$

To solve it, we get

$$P_0(t) = C_1 e^{k_1 t} + C_2 e^{k_2 t} \tag{36}$$

where,

$$k_1 = \frac{-(n\lambda\omega_{00} + \mu\omega_{11}) - \sqrt{(n\lambda\omega_{00} - \mu\omega_{11})^2 + 4n\lambda\mu\omega_{01}\omega_{10}}}{2} \tag{37}$$

$$k_2 = \frac{-(n\lambda\omega_{00} + \mu\omega_{11}) + \sqrt{(n\lambda\omega_{00} - \mu\omega_{11})^2 + 4n\lambda\mu\omega_{01}\omega_{10}}}{2} \tag{38}$$

Substituting equation Eq. 36 in the first equation of equation system Eq. 34, we get

$$P_1(t) = \frac{k_1 C_1 + n\lambda C_1 \omega_{00}}{\omega_{10}\mu} e^{k_1 t} + \frac{k_2 C_2 + n\lambda C_2 \omega_{00}}{\omega_{10}\mu} e^{k_2 t} \tag{39}$$

By substituting the initial conditions in Eq. 36 and Eq. 39, we can get

$$C_1 = \frac{-k_2 - n\lambda\omega_{00}}{k_1 - k_2} \tag{40}$$

$$C_2 = \frac{k_1 + n\lambda\omega_{00}}{k_1 - k_2} \tag{41}$$

The fuzzy validity of the system is

$$A(t) = \frac{-k_2 - n\lambda\omega_{00}}{k_1 - k_2} e^{k_1 t} + \frac{k_1 + n\lambda\omega_{00}}{k_1 - k_2} e^{k_2 t} \tag{42}$$

Make the fault state "1" of the system as an absorption state, that is $\mu = 0$ (when the system fails, it will not be repaired). In this way, a new fuzzy Markov process $\{Y(t), t \geq 0\}$ is formed

$$R(t) = P_0(t) = P\{Y(t) = 0\} \tag{43}$$

Similar to the previous derivation, the equations are obtained

$$P_0(t+\Delta t) = P_0(t)P_{00}(\Delta t) = \omega_{00}(1-n\lambda\Delta t)P_0(t) + 0(\Delta t) \tag{44}$$

$$P_1(t+\Delta t) = P_0(t)P_{01}(\Delta t) + P_1(t)P_{11}(\Delta t) = n\lambda\omega_{01}P_0(t)\Delta t + \omega_{11}P_1(t) + 0(\Delta t) \tag{45}$$

When $\Delta t \to 0$, the differential equations and the initial conditions are

$$\begin{cases} P'_0(t) = -n\lambda\omega_{00}P_0(t) \\ P'_1(t) = n\lambda\omega_{01}P_0(t) \\ P_0(0) = 1, P_1(0) = 0 \end{cases} \tag{46}$$

The solutions of differential equations (46) satisfying the initial conditions are as follows

$$P_0(t) = e^{-n\lambda\omega_{00}t} \tag{47}$$

$$P_1(t) = -\frac{\omega_{01}}{\omega_{00}}e^{-n\lambda\omega_{00}t} + \frac{\omega_{01}}{\omega_{00}} \tag{48}$$

$$R(t) = e^{-n\lambda\omega_{00}t} \tag{49}$$

Repairable Series System with Different n Components

Suppose that the distribution function of the life $T_1, T_2, \cdots, T_n$ of different n components is

$$P(T_i \le t) = 1 - e^{-\lambda_i t} \quad (\lambda_i > 0, \quad i = 1, \ 2, \ \ldots, \ n) \tag{50}$$

The distribution function of repair time τ_i after failure is

$$M(\tau_i \le t) = 1 - e^{-\mu_i t} \quad (\mu_i > 0, \quad i = 1, \ 2, \ \ldots, \ n) \tag{51}$$

When a component in the system fails, the system fails, and a repairman immediately repairs the failed component.

Similar to the previous derivation, the fuzzy validity of the system can be obtained as follows:

$$A_1(t) = \frac{-k_2 - (\lambda_1 + \lambda_2 + \cdots + \lambda_n)\omega_{00}}{k_1 - k_2} e^{k_1 t}$$

$$+ \frac{k_1 + (\lambda_1 + \lambda_2 + \cdots + \lambda_n)\omega_{00}}{k_1 - k_2} e^{k_2 t} \tag{52}$$

where,

$$k_1, k_2 = \frac{-((\lambda_1 + \lambda_2 + \cdots + \lambda_n)\omega_{00} + \mu_i \omega_{11}) \mp}{}$$

$$\frac{\sqrt{((\lambda_1 + \lambda_2 + \cdots + \lambda_n)\omega_{00} - \mu_i \omega_{11})^2 + 4\mu_i \omega_{01} \omega_{10} (\lambda_1 + \lambda_2 + \cdots + \lambda_n)}}{2} \tag{53}$$

The fuzzy reliability of the system is

$$R_1(t) = e^{-\omega_{00}(\lambda_1 + \lambda_2 + \cdots + \lambda_n)t} \tag{54}$$

The fuzzy reliability of repairable series system composed of n components and several repairmen can be studied by reference to this section.

Fuzzy Comprehensive Evaluation Method of Reliability Distribution

According to complexity, technical development level, working time and environmental conditions, scoring distribution method scores and assigns reliability to each subsystem scoring conditions. There are only four factors considered in the method of the score distribution, which are not comprehensive enough. Each reason is not divided into different grades, and a fuzzy comprehensive evaluation is not carried out. In this paper, the two-level fuzzy comprehensive evaluation method is used to evaluate each sub-system, calculate the scoring coefficient, and assign reliability to each sub-system.

Fuzzy Comprehensive Evaluation of Subsystem

Build factor set:

Design level, manufacturing level, material quality, importance degree, complexity, working time, environmental conditions and maintenance cost are the factors that affect the subsystem

Each factor is subdivided into several levels according to its nature and degree. For example, the factor level set of design level can be represented by a set (very high, high, high, general, low, low, very low).

Establishment of an alternative set:

Since the scoring coefficient C is limited to the value of [0, 1], the interval of [0, 1] is discretized in equal steps, and the set of discrete values C_k (k=1, 2, …, n) are taken

$$C = \{C_1, \ C_2, \ …, \ C_n\} \tag{55}$$

As an alternative set.

First level fuzzy comprehensive evaluation:

The first level fuzzy comprehensive evaluation is actually a kind of single-factor evaluation, which is made to deal with the fuzziest of factors by integrating the contribution of each level of a cause to the value of evaluation object. The membership degree of each grade evaluation set is a matrix composed of rows.

$$\underset{\sim}{R}_i = \begin{pmatrix} r_{i11} & r_{i12} & \cdots & r_{i1n} \\ r_{i21} & r_{i22} & \cdots & r_{i2n} \\ \cdots & \cdots & \cdots & \cdots \\ r_{im1} & r_{im2} & \cdots & r_{imn} \end{pmatrix} \tag{56}$$

It is called the rating matrix of the ith factor.

If the rank order of each cause is arranged in accordance with the trend of its influence on the value of the evaluation object, the same rank evaluation matrix of each cause can be obtained.

In order to show the influence of a certain factor on the value of the evaluation object, we should give proper weight to each level of the factor, that is to say, we should set up the corresponding level weight set for each factor. The membership degree $\mu_{ij}(i=1,2,\cdots,8; j=1,2, \ldots, m)$ of the j grade of the I factor to the factor is normalized

$$a_{ij} = \mu_{ij} / \sum_{j=1}^{m} \mu_{ij} \tag{57}$$

Taking as the weight of the level, the grade weight set of the ith factor is obtained as follows:

$$\underset{\sim}{A}_i = (a_{i1}, \ a_{i2}, \ \ldots, \ a_{im}) \tag{58}$$

According to the fuzzy subset of each level of the ith factor, the comprehensive evaluation is carried out. The first level fuzzy comprehensive evaluation set is as follows:

$$\underset{\sim}{B}_i = \underset{\sim}{A}_i \circ \underset{\sim}{R}_i$$

$$= (a_{i1},\ a_{i2},\ \ldots,\ a_{im}) \circ \begin{pmatrix} r_{i11} & r_{i12} & \cdots & r_{i1n} \\ r_{i21} & r_{i22} & \cdots & r_{i2n} \\ \cdots & \cdots & \cdots & \cdots \\ r_{im1} & r_{im2} & \cdots & r_{imn} \end{pmatrix}$$

$$= (b_{i1},\ b_{i2},\ \ldots,\ b_{in}) \tag{59}$$

Using the comprehensive evaluation model IV,. $M(\cdot,\ +)$ is calculated as follows:

$$b_{ik} = \sum_{j=1}^{m} a_{ij} r_{ijk} \tag{60}$$

$i=1,2,\ldots,\ 8\ ;\ j=1,\ 2,\ \ldots,\ m\ ;\ k=1,\ 2,\ \ldots,\ n)$

A matrix composed of elements b_{ik} is given by

$$\underset{\sim}{R} = \begin{pmatrix} \underset{\sim}{B_1} \\ \underset{\sim}{B_2} \\ \vdots \\ \underset{\sim}{B_8} \end{pmatrix} = \begin{pmatrix} b_{11} & b_{12} & \cdots & b_{1n} \\ b_{21} & b_{22} & \cdots & b_{2n} \\ \cdots & \cdots & \cdots & \cdots \\ b_{81} & b_{82} & \cdots & b_{8n} \end{pmatrix} \tag{61}$$

It is called the level comprehensive evaluation matrix, that is, the first level fuzzy comprehensive evaluation matrix.

Two level fuzzy comprehensive evaluation:

The first level fuzzy comprehensive evaluation integrates the contribution of each level and reflects the influence of a factor on the value of the scoring coefficient. Therefore, the matrix $\underset{\sim}{R}$ is a single factor evaluation matrix of the two-level fuzzy comprehensive evaluation.

Generally, the importance of each factor influencing the value of the scoring coefficient is different. Let a_i be the *ith* (i=1, 2, …, 8) weight of factors, then the factor weight set reflecting the importance of each factor is given as:

$$\underset{\sim}{A} = (a_1, \ a_2, \ …, \ a_8) \tag{62}$$

All weights shall meet the conditions of oneness and nonnegativity.

Then, according to the comprehensive evaluation of all the influencing factors, the two-level fuzzy comprehensive evaluation set is as follows:

$$\underset{\sim}{B} = \underset{\sim}{A} \cdot \underset{\sim}{R}$$

$$= (a_1, \ a_2, \ …, \ a_8) \cdot \begin{pmatrix} b_{11} & b_{12} & … & b_{1n} \\ b_{21} & b_{22} & … & b_{2n} \\ … & … & … & … \\ b_{81} & b_{82} & … & b_{8n} \end{pmatrix} \tag{63}$$

The above formula is still calculated by ordinary matrix multiplication

$$\underset{\sim}{B} = (b_1, \ b_2, \ …, \ b_n) \tag{64}$$

where,

$$b_k = \sum_{i=1}^{8} a_i b_{ik} \qquad (k=1, \ 2, \ …, \ n) \tag{65}$$

Finally, the scoring coefficient determined by the weighted average method is as follows:

$$C = \frac{\sum_{k=1}^{n} b_k C_k}{\sum_{k=1}^{n} b_k} \tag{66}$$

Series System

The system is composed of d units in series, the reliability of unit A_i (i=1, 2, ..., d) is R_i. The expected reliability of this unit under the consideration of scoring coefficient C_i is

$$R^*_i = 1 - C_i(1 - R_i) \tag{67}$$

The reliability index of the system is required to be R_s and distributed to d units according to the equal distribution principle, then the reliability of each unit is $\sqrt[d]{R_s}$. By equation

$$\sqrt[d]{R_s} = 1 - C_i(1 - R_i) \tag{68}$$

It can be found that the reliability distribution value of the ith unit is

$$R_i = 1 - \frac{1 - \sqrt[d]{R_s}}{C_i} \qquad (i\text{=}1, \ 2, \ ..., \ d) \tag{69}$$

Parallel System

The system is composed of e units in parallel, If the reliability of unit B_j (j=1, 2, ..., e) is R_j, the expected reliability of this unit under the consideration of scoring coefficient C_j is

$$R^*_j = 1 - C_j(1 - R_j) \tag{70}$$

The reliability index of the system is required to be R_t and allocated to e units according to the principle of equal distribution, then the reliability of each unit is $1 - (1 - R_t)^{\frac{1}{e}}$. By equation

$$1 - C_j(1 - R_j) = 1 - (1 - R_t)^{\frac{1}{e}} \tag{71}$$

From the above equation, we can get the reliability distribution value of the jth element as

$$R_j = 1 - \frac{(1-R_t)^{\frac{1}{e}}}{C_j} \qquad (j=1, \ 2, \ \ldots, \ e) \qquad (72)$$

Fuzzy Method of Reliability Distribution

The common methods of reliability allocation are equal distribution method, redistribution method, proportion distribution method, comprehensive scoring distribution method, and dynamic planning distribution method. Fuzzy factors, such as design level, manufacturing level, material quality, importance, complexity, working time, environmental conditions, maintenance cost, *etc.*, have an impact on the system, and the fuzzy method shall be adopted for system reliability distribution. The fuzzy comprehensive evaluation method of reliability distribution uses the two-level fuzzy comprehensive evaluation method to evaluate each subsystem, find the scoring coefficient, and assign reliability to each subsystem. The fuzzy optimization method of reliability allocation is to minimize the objective function (cost or weight, or size, in this paper, cost is taken as the objective function) and meet the constraints of reliability index. Fuzzy comprehensive evaluation method for reliability distribution of the hybrid system.

The hybrid system shown in Fig. (**1**) is transformed into an equal system first, as shown in Fig. (**2**) and Fig. (**3**).

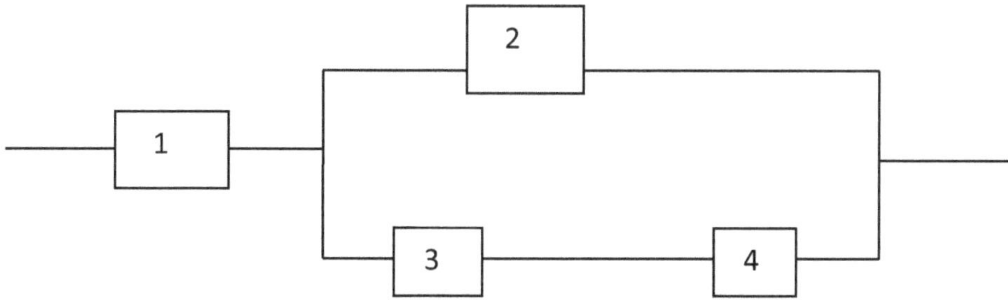

Fig. (1). The hybrid system.

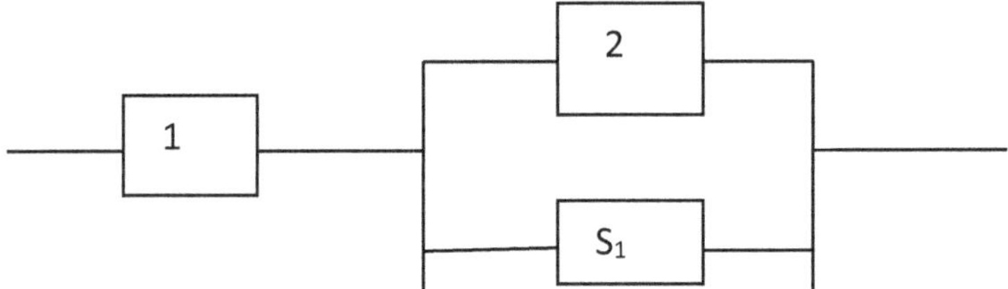

Fig. (2). Equivalent system.

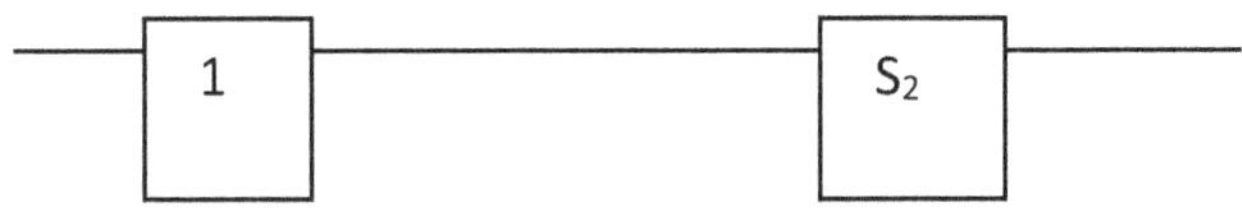

Fig. (3). Equivalent system.

For Fig. (**3**), the reliability allocated to each unit according to formula (69) is

$$R_1 = 1 - \frac{1 - \sqrt{R_s}}{C_1} \tag{73}$$

$$R_{s_2} = 1 - \frac{1 - \sqrt{R_S}}{C_{s_2}} \tag{74}$$

For Fig. (**3**), the reliability allocated to each unit according to formula (72) is

$$R_2 = 1 - \frac{1 - \sqrt{R_{s_2}}}{C_2} \tag{75}$$

$$R_{s_1} = 1 - \frac{1 - \sqrt{R_{s_2}}}{C_{s_1}} \tag{76}$$

For Fig. (**2**), the reliability allocated to each unit according to formula (69) is

$$R_3 = 1 - \frac{1 - \sqrt{R_{S_1}}}{C_3} \tag{77}$$

$$R_4 = 1 - \frac{1 - \sqrt{R_{S_1}}}{C_4} \tag{78}$$

Fuzzy Optimization Method

It is difficult to give the lower bound of reliability very accurately. Because its value is affected by many fuzzy factors, such as design level, manufacturing level, material quality, use conditions, *etc.*, in fact, it should be considered intermediate state from the completely forbidden state to the fully permitted state, which is regarded as the fuzzy subset in the design space.

Fuzzy optimization model:

Series System

objective function

$$\min \sum_{i=1}^{n} G_i(\overset{\vee}{R}_i, \ R_i) \tag{79}$$

constraint conditions

$$\prod_{i=1}^{n} R_i \geqslant R_{\underset{\sim}{S_L}} \tag{80}$$

$$\hat{R}_{\underset{\sim}{i_L}} \leqslant R_i \quad i = 1, 2, \cdots, \ n \tag{81}$$

where: $R_{\underset{\sim}{S_L}}$ ——Lower bound value of specified system reliability index;

R_i——Reliability assigned by unit i;

$\underset{\sim}{\overset{\vee}{R}}_{i_L}$——Unit i lower bound of existing predicted reliability;

$G_i\left(\overset{\vee}{R}_i,\ R_i\right)$——The cost function of the ith unit reliability from $\overset{\vee}{R}_i$ to R_i

$\overset{\vee}{R}_i$——Unit i reliability of existing predictions, $\quad \overset{\vee}{R}_i = \underset{\sim}{\overset{\vee}{R}}_{i_L}$。

Parallel System

objective function

$$\min \sum_{i=1}^{n} G_i(\overset{\vee}{F}_i, F_i) \tag{82}$$

constraint conditions

$$\prod_{i=1}^{n} F_i \leqslant \underset{\sim}{F}_{s_U} \tag{83}$$

$$F_i \leqslant \underset{\sim}{\overset{\vee}{F}}_{i_U} \tag{84}$$

Where: $\underset{\sim}{F}_{s_U}$——Upper bound value of specified system unreliability;

F_i——Unit assigned unreliability

$\underset{\sim}{\overset{\vee}{F}}_{i_U}$——The upper bound value of the predicted value of the unreliability of unit i;

$\check{F_i}$——Prediction value of unreliability of unit i, $\check{F_i} = \check{F}_{\sim i_U}$;

$G_i(\check{F_i},\ F_i)$——The cost function of the unreliability of unit i is reduced from $\check{F_i}$ to F_i .

Hybrid System

objective function

$$\min \sum_{i=1}^{n} G_i(\hat{R_i},\ R_i) \tag{85}$$

constraint conditions

$$f(R_i) \geqslant R_{\sim t_L} \tag{86}$$

$$\hat{R}_{\sim i_L} \leq R_i \tag{87}$$

where: $R_{\sim t_L}$——Lower bound value of specified system reliability index;

R_i——Reliability assigned by unit i ;

$\hat{R}_{\sim i_L}$——Unit i lower bound of existing predicted reliability;

$\hat{R_i}$——Prediction value of the reliability of unit i , $\hat{R_i} = \hat{R}_{\sim i_L}$;

$F(R_i)$——Reliability of hybrid system;

$G(\hat{R}_i, R_i)$——The cost function of the reliability of unit i is reduced from $\hat{R}_i$ to R_i。

Fuzzy Optimization Method for Series System

The optimal level cut set method is a very effective and simple method to solve the fuzzy optimization problem. Its basic idea is to find an optimal level cut set, corresponding to the optimal solution of the cut set, which is the optimal solution of the original fuzzy optimization problem.

In order to simplify the calculation, the fuzzy constrained membership function adopts the linear membership function. By using the optimal level cut set method, the fuzzy constraints can be transformed into common constraints as follows:

$$\prod_{i=1}^{n} R_i \geqslant \underline{R}_S^L + (\underline{R}_s^u - \underline{R}_s^L)\lambda^* \tag{88}$$

$$R_i \geqslant \underline{\overset{\vee}{R}}_i^L + (\underline{\overset{\vee}{R}}_i^u - \underline{\overset{\vee}{R}}_i^L)\lambda^*, \quad i = 1,2, \; \ldots, \; n \tag{89}$$

$\underline{R}_s^L, \; \underline{R}_s^u, \underline{\overset{\vee}{R}}_i^u, \underline{\overset{\vee}{R}}_i^L$ are by the amplification coefficient method.

It is determined by a two-level fuzzy comprehensive evaluation. In the fuzzy comprehensive evaluation method, each factor that affects the value is divided into several grades according to its nature and degree, and each factor and each grade are regarded as a fuzzy subset in the hierarchy field. Then, a first level fuzzy comprehensive evaluation is carried out according to each grade fuzzy subset, and the evaluation result is regarded as a single factor evaluation set, and then a two - level fuzzy comprehensive evaluation is carried out according to all factors to determine λ^*. The fuzzy optimization problem of reliability allocation for parallel and hybrid systems can be solved with reference to the above.

CONCLUDING REMARKS

Considering the influence of fuzzy factors, the fuzzy reliability calculation formula of a single component maintenance system and the repairable series system is given. Using the theory of fuzzy random process to study the reliability of the maintenance system, considering the fuzziness of the system, it is more reasonable and more practical. Two fuzzy methods for reliability allocation are proposed. The fuzzy method of reliability allocation considers the influence of fuzzy factors on the system. It is more scientific and practical than the conventional method.

CONSENT FOR PUBLICATION

Not applicable.

CONFLICT OF INTEREST

The author declares no conflict of interest, financial or otherwise.

ACKNOWLEDGEMENTS

Declared none.

REFERENCES

[1]   L.V. Utkin, and S.V. Gurov, "A general formal approach for fuzzy reliability analysis in the possibility context", *Fuzzy Sets Syst.,* vol. 83, pp. 203-213, 1996.
[http://dx.doi.org/10.1016/0165-0114(95)00391-6]

[2]   R.S. Chanda, and P.K. Bhattacharjee, "A reliability approach to transmission expansion planning using fuzzy fault-tree model", *Electr. Power Syst. Res.,* vol. 45, pp. 101-108, 1998.
[http://dx.doi.org/10.1016/S0378-7796(97)01226-1]

[3]   G.R. Dodagoudar, "G. Venkatachalam, "Reliability analysis of slopes using fuzzy sets theory", *Comput. Geotech.,* vol. 27, pp. 101-115, 2000.
[http://dx.doi.org/10.1016/S0266-352X(00)00009-4]

[4]   M. Savoia, "Structural reliability analysis through fuzzy number approach, with application to stability", *Comput. Struc.,* vol. 80, pp. 1087-1102, 2002.
[http://dx.doi.org/10.1016/S0045-7949(02)00068-8]

[5]   E. Zio, and P. Baraldi, "Sensitivity analysis and fuzzy modelling for passive systems reliability assessment", *Ann. Nucl. Energy,* vol. 31, pp. 277-301, 2004.
[http://dx.doi.org/10.1016/S0306-4549(03)00220-2]

[6]   G.S. Mahapatra, and T.K. Roy, *"Fuzzy multi-objective mathematical programming on reliability optimization model", Applied Mathematics and Computation,* vol. 174. January, 2006, pp. 643-659.

[7]   A. Bagis, "Fuzzy rule base design using tabu search algorithm for nonlinear system modeling", *ISA Trans.,* vol. 47, no. 1, pp. 32-44, 2008.
[http://dx.doi.org/10.1016/j.isatra.2007.09.001] [PMID: 17945233]

[8]   L. Görkemli, *Selda Kapan Ulusoy, "Fuzzy Bayesian reliability and availability analysis of production systems", Computers & Industrial Engineering.,* vol. 59. December, 2010, pp. 690-696.

[9]   M. Kumar, and S.P. Yadav, Mohit Kumar Shiv, "A novel approach for analyzing fuzzy system reliability using different types of intuitionistic fuzzy failure rates of components", *ISA Trans.,* vol. 51, no. 2, pp. 288-297, 2012.
[http://dx.doi.org/10.1016/j.isatra.2011.10.002] [PMID: 22134065]

[10]   "Julwan, Hendry Purba, "A fuzzy-based reliability approach to evaluate basic events of fault tree analysis for nuclear power plant probabilistic safety assessment", *Ann. Nucl. Energy,* vol. 70, pp. 21-29, 2014.
[http://dx.doi.org/10.1016/j.anucene.2014.02.022]

[11]   O. Hryniewicz, *"Bayes statistical decisions with random fuzzy data—an application in reliability",Reliability Engineering & System Safety.,* vol. 151. Junuary, 2016, pp. 20-33.

[12]   D.S. González-González, R.J. Praga-Alejo, and M. Cantú-Sifuentes, "A non-linear fuzzy degradation model for estimating reliability of a polymeric coating", *Appl. Math. Model.,* vol. 40, pp. 1387-1401, 2016.
[http://dx.doi.org/10.1016/j.apm.2015.06.033]

[13]   Sayed Javad Aghili, "Hamze Hajian-Hoseinabadi, "Reliability evaluation of repairable systems using various fuzzy-based methods – A substation automation case study", *Int. J. Electr. Power Energy Syst.,* vol. 85, pp. 130-142, 2017.
[http://dx.doi.org/10.1016/j.ijepes.2016.08.010]

[14]   ChongWang, Hermann G.Matthies, *"Epistemic uncertainty-based reliability analysis for engineering system with hybrid evidence and fuzzy variables". Computer Methods in Applied Mechanics and Engineering.,* vol. 355. Junuary, 2019, pp. 438-455.

CHAPTER 2

Reliability Calculation

Abstract: Considering the influence of random factors, such as material inhomogeneity and manufacturing and installation error, the reliability of rigid rotor balance is studied by using stress -strength interference model. Based on the sensitivity analysis, a Monte Carlo simulation for the reliability calculation of gears is proposed. Based on sensitivity analysis, an optimization method for reliability calculation is proposed. The reliability calculation of spring is studied by the HL-RF method.

Keywords: Balance, Gear, HL-RF method, Monte Carlo simulation, Optimization method, Random factors, Rigid rotor, Reliability calculation, Sensitivity analysis, Spring.

INSTRUCTION

In the reliability design, some variables in the conventional design, such as load, material strength, geometric dimension of parts, *etc.*, are treated as random variables. The data used in the design comes from the test or practice. After statistical analysis, the change in working conditions and the influence of various random factors are considered. The reliability design makes the designer understand the failure probability. The designed product is lighter than the safety factor method. The reliability design saves the raw materials, improves product quality, and makes the product more competitive.

An algorithm HL-RF for the calculation of structural reliability under combined loading is formulated [1]. The first-order reliability method [2, 4] and the second -order reliability method (SORM) [3] are proposed. The importance of sample as a special technology in reliability calculation is a highly efficient method [5]. The reliability analysis of a 75 m tall steel lattice tower is presented for the annual variation of mean wind velocity taken as a function of both mean wind speed and direction [6]. Directional simulation in the load space is a simulation-based technique used to perform the reliability analysis of structures [7]. New imprecise structural reliability models are developed based on the imprecise Bayesian inference and are imprecise Dirichlet distribution, imprecise negative binomial, gamma-exponential, and normal models [8].

Wenhui Mo

Alternative methods for constructing environmental contours for probabilistic structural reliability analysis of structures are proposed [9]. An integrated model for the reliability analysis of failure data is presented jointly with a robust structural reliability model [10]. The analysis of the probability of failure of a load-bearing steel bridge member under bending is proposed [11]. A 5-layer BPN is proposed for the reliability analysis of structural systems [12].

The reliability of static and dynamic balance of rigid rotor is discussed. Based on sensitivity analysis, the reliability calculation of contact fatigue strength and bending fatigue strength of gears is studied by Monte Carlo simulation. The reliability is calculated by the optimization method. The multi-objective optimization model can be established for the reliability calculation of various failure modes, which can be obtained by the multi-objective optimization method. The reliability design of spring is studied by the HL-RF method.

RELIABILITY OF RIGID ROTOR BALANCE

The balance of a rigid rotor can be divided into static balance and dynamic balance. Due to the influence of the random factors such as the uneven material and the manufacturing and installation errors, the central inertia principal axis of all the actual rotors deviates more or less from its rotation axis. Unbalance will cause the deflection and internal stress of the rotor, make the machine produce vibration and noise, accelerate the wear of the bearing and other parts, reduce the working efficiency of the machine, and even cause various accidents in serious cases. Therefore, it is necessary to study the reliability of rotor balance

RELIABILITY CALCULATION OF STATIC BALANCE

Unbalanced mass in the same rotating surface is $m_i \pm \Delta m_i$, i=1, 2, …, j. Their vector diameters are $r_{i\,a_i}^{\,b_i}$, b_i, a_i are the upper and lower deviations of r_i, and the angle between r_i and the horizontal axis is $\alpha_i \pm \Delta \alpha_i$. It is known that the added balance mass $m_p \pm \Delta m_p$ and its vector diameter $r_{p\,c}^{\,d}$, , d and c are the upper and lower deviations of r_{p}, and the angle between r_{p} and the horizontal coordinate axis is $\alpha_p \pm \Delta \alpha_p$. The rotor speed is n $\pm \Delta$ n. Their mean and standard deviation are:

$$u_{m_i} = m_i \qquad \sigma_{m_i} = \frac{\Delta m_i}{3}$$

$$u_{\alpha_i} = \alpha_i \qquad \sigma_{\alpha_i} = \frac{\Delta \alpha_i}{3}$$

$$u_{\alpha_p} = \alpha_p \qquad \sigma_{\alpha_p} = \frac{\Delta \alpha_p}{3}$$

$$u_{r_i} = \frac{2r_i + a_i + b_i}{2} \qquad \sigma_{r_i} = \frac{b_i - a_i}{6}$$

$$u_{m_p} = m_p \qquad \sigma_{m_p} = \frac{\Delta m_p}{3}$$

$$u_{r_p} = \frac{2r_p + c + d}{2} \qquad \sigma_{r_p} = \frac{d - c}{6}$$

Let m and r be the total mass and vector diameter of the total center of mass of the rotor, and mr be the mass-radius product. The mass-radius product expresses the magnitude and direction of the centrifugal force of each mass at the same speed. Thus, we obtain

$$mr = m_p r_p + \sum_{i=1}^{j} m_i r_i \tag{1}$$

Eq. 1 is decomposed along the horizontal coordinate axis and the vertical coordinate axis

$$mr_t = m_p r_p \cos \alpha_p + \sum_{i=1}^{j} m_i r_i \cos \alpha_i \tag{2}$$

$$mr_s = m_p r_p \sin \alpha_p + \sum_{i=1}^{j} m_i r_i \sin \alpha_i \tag{3}$$

Scalar r_i refers to the size of the vector of the unbalanced mass m_i ($i=1$, 2, ..., j) and r_p refers to the size of the vector of the added balanced mass

From Eq. 2 and Eq. 3, we obtain:

$$r_t = \frac{m_\mathrm{p} r_\mathrm{p} \cos \alpha_\mathrm{p}}{m} + \sum_{i=1}^{j} \frac{m_i r_i \cos \alpha_i}{m} \tag{4}$$

$$r_s = \frac{m_\mathrm{p} r_\mathrm{p} \sin \alpha_\mathrm{p}}{m} + \sum_{i=1}^{j} \frac{m_i r_i \sin \alpha_i}{m} \tag{5}$$

The size of r is

$$r = \sqrt{r_t^2 + r_s^2} \tag{6}$$

Because $m = m_\mathrm{p} + \sum_{i=1}^{j} m_i$ $\tag{7}$

Therefore $\mu_m = \mu_{m_\mathrm{p}} + \sum_{i=1}^{j} \mu_{m_i}; \sigma_m^2 = \sigma_{m_\mathrm{p}}^2 + \sum_{i=1}^{j} \sigma_{m_i}^2$ $\tag{8}$

The mean and standard deviation of r_t, r_s, r are given as follows:

$$\mu_{r_t} = \frac{\mu_{m_\mathrm{p}} \mu_{r_\mathrm{p}} \cos \mu_{\alpha_\mathrm{p}}}{\mu_m} + \sum_{i=1}^{j} \frac{\mu_{m_i} \mu_{r_i} \cos \mu_{\alpha_i}}{\mu_m} \tag{9}$$

$$\mu_{r_s} = \frac{\mu_{m_\mathrm{p}} \mu_{r_\mathrm{p}} \sin \mu_{\alpha_\mathrm{p}}}{\mu_m} + \sum_{i=1}^{j} \frac{\mu_{m_i} \mu_{r_i} \sin \mu_{\alpha_i}}{\mu_m} \tag{10}$$

$$\mu_r = \sqrt{\mu_{r_t}^2 + \mu_{r_s}^2} \tag{11}$$

$$\sigma_{r_t}^2 = \left(\frac{\mu_{r_\mathrm{p}} \cos \mu_{\alpha_\mathrm{p}}}{\mu_m} \right)^2 \sigma_{m_\mathrm{p}}^2 + \left(\frac{\mu_{m_\mathrm{p}} \cos \mu_{\alpha_\mathrm{p}}}{\mu_m} \right)^2 \sigma_{r_\mathrm{p}}^2$$

$$+\left(\frac{\mu_{m_p}\mu_{r_p}\sin\mu_{\alpha_p}}{\mu_m}\right)^2\sigma_{\alpha_p}^2+$$

$$\sum_{i=1}^{j}\left[\begin{array}{c}\left(\dfrac{\mu_{m_i}\cos\mu_{\alpha_i}}{\mu_m}\right)^2\sigma_{r_i}^2+\left(\dfrac{\mu_{r_i}\cos\mu_{\alpha_i}}{\mu_m}\right)^2\sigma_{m_i}^2+\\[4mm]\left(\dfrac{\mu_{m_i}\mu_{r_i}\sin\mu_{\alpha_i}}{\mu_m}\right)^2\sigma_{\alpha_i}^2\end{array}\right]$$

$$+\frac{(\mu_{m_p}\mu_{r_p}\cos\mu_{\alpha_p}+\sum\limits_{i=1}^{j}\mu_{m_i}\mu_{r_i}\cos\mu_{\alpha_i})^2}{\mu_m^4}\sigma_m^2 \tag{12}$$

$$\sigma_{r_s}^2=\left(\frac{\mu_{r_p}\sin\mu_{\alpha_p}}{\mu_m}\right)^2\sigma_{m_p}^2+\left(\frac{\mu_{m_p}\sin\mu_{\alpha_p}}{\mu_m}\right)^2\sigma_{r_p}^2+$$

$$\left(\frac{\mu_{m_p}\mu_{r_p}\cos\mu_{\alpha_p}}{\mu_m}\right)^2\sigma_{\alpha_p}^2+$$

$$\sum_{i=1}^{j}\left[\begin{array}{c}\left(\dfrac{\mu_{r_i}\sin\mu_{\alpha_i}}{\mu_m}\right)^2\sigma_{m_i}^2+\left(\dfrac{\mu_{m_i}\sin\mu_{\alpha_i}}{\mu_m}\right)^2\sigma_{r_i}^2+\\[4mm]\left(\dfrac{\mu_{m_i}\mu_{r_i}\cos\mu_{\alpha_i}}{\mu_m}\right)^2\sigma_{\alpha_i}^2\end{array}\right]$$

$$+\frac{(\mu_{m_p}\mu_{r_p}\sin\mu_{\alpha_p}+\sum\limits_{i=1}^{j}\mu_{m_i}\mu_{r_i}\sin\mu_{\alpha_i})^2}{\mu_m^4}\sigma_m^2 \tag{13}$$

$$\sigma_r^2=\frac{\mu_{r_t}^2}{\mu_{r_t}^2+\mu_{r_s}^2}\sigma_{r_t}^2+\frac{\mu_{r_s}^2}{\mu_{r_t}^2+\mu_{r_s}^2}\sigma_{r_s}^2 \tag{14}$$

According to the specific type of rotor, check the standard of "balance precision of rigid rotor", the value A of balance precision grade can be obtained as follows:

$$\mu_{\mathrm{n}} = n \qquad \sigma_{\mathrm{n}} = \frac{\Delta n}{3}$$

$$\mu_{\omega} = \frac{\pi \mu_{\mathrm{n}}}{30} \qquad \sigma_{\omega}^2 = \frac{\pi^2}{900}\sigma_{\mathrm{n}}^2$$

and $(r) = \dfrac{1000A}{\omega}$

The mean and standard deviation of the allowable unbalance (r) of the rotor are as follows:

$$\mu_{(r)} = \frac{1000A}{\mu_{\omega}} \tag{15}$$

$$\sigma_{(r)}^2 = \frac{1000000A^2}{\mu_{\omega}^4}\sigma_{\omega}^2 \tag{16}$$

The reliability of static balance is given as:

$$R_1 = \Phi\left[\frac{\mu_{(\mathrm{r})} - \mu_{\mathrm{r}}}{\sqrt{\sigma_{(\mathrm{r})}^2 + \sigma_{\mathrm{r}}^2}}\right] \tag{17}$$

RELIABILITY CALCULATION OF DYNAMIC BALANCE

Suppose that the unbalance mass of the rotor is distributed in rotating surfaces 1, 2, ..., k, expressed as $m'_i \pm \Delta m'_i$, i=1, 2, ..., k. Their rotation radius vector $r_{ia'_i}^{b'_i}$, b'_i, a'_i are upper and lower deviation of r_i. m'_i can be replaced by the sum of the other two masses m''_{i_1} and m''_{i_2} in any two parallel planes (the plane of revolution). The distance between the plane T' and the plane T'' is $l_{l_2}^{l_1}$.

The distance between plane i where m'_i is located and the plane T' is $l'_{i\,l'_{i_2}}^{\,l'_{i_1}}$.
The distance between plane where m'_i is located and plane T'' is $l''_{i\,l''_{i_2}}^{\,l''_{i_1}}$, l_1, l_2, l'_{i_1}, l'_{i_2}, l''_{i_1}, l''_{i_2} are their upper and lower deviations, then

$$m''_{i_1} = \frac{l''_i}{l}m'_i \qquad m''_{i_2} = \frac{l'_i}{l}m'_i$$

Their mean and standard deviation are

$$\mu_{m'_i} = m'_i \qquad \sigma_{m'_i} = \frac{\Delta m'_i}{3}$$

$$\mu_l = \frac{2l + l_1 + l_2}{2}$$

$$\sigma_l = \frac{l_1 - l_2}{6}$$

$$\mu_{l'_i} = \frac{2l'_i + l'_{i1} + l'_{i2}}{2}$$

$$\sigma_{l'_i} = \frac{l'_{i_1} - l'_{i_2}}{6}$$

$$\mu_{l''_i} = \frac{2l''_i + l''_{i_1} + l''_{i_2}}{2}$$

$$\sigma_{l''_i} = \frac{l''_{i_1} - l''_{i_2}}{6}$$

$$\mu_{m''_{i_1}} = \frac{\mu_{l''_i}}{\mu_l}\mu_{m'_i} \qquad \mu_{m''_{i_2}} = \frac{\mu_{l'_i}}{\mu_l}\mu_{m'_i}$$

$$\sigma^2_{m''_{i_1}} = \left(\frac{\mu_{m'_i}}{\mu_l}\right)^2 \sigma^2_{l''_i} + \frac{\mu^2_{l''_i}\mu^2_{m'_i}}{\mu^4_l}\sigma^2_l + \left(\frac{\mu^2_{l''_i}}{\mu_l}\right)^2 \sigma^2_{m'_i} \qquad\qquad (18)$$

$$\sigma^2_{\mathrm{m''}_{i_2}} = \left(\frac{\mu_{\mathrm{m'}_i}}{\mu_l}\right)^2 \sigma^2_{l'_i} + \frac{\mu^2_{l'_i}\mu^2_{\mathrm{m'}_i}}{\mu^4_l}\sigma^2_l + \left(\frac{\mu_{\mathrm{m'}_i}}{\mu_l}\right)^2 \sigma^2_{\mathrm{m'}_i} \tag{19}$$

In the rotating surface T', if the added balance mass is $m'_{\mathrm{p}} \pm \Delta m'_{\mathrm{p}}$ and the radius vector is r'_{p}, then

$$m'r_1 = m'_{\mathrm{p}}r'_{\mathrm{p}} + \sum_{i=1}^{k} m''_{i_1}r'_i \tag{20}$$

μ_{r_1} can be obtained from Eq. 11, $\sigma^2_{r_1}$ can be obtained from Eq. 14, $\mu_{(r_1)}$ can be obtained from Eq. 15, $\sigma^2_{(r_1)}$ can be obtained from Eq. 16, $R_{T'}$ can be obtained from Eq. 17, and $R_{T'}$ is the reliability of balance in the rotating surface T'.

In the rotating surface T'', the added balance mass is $m''_{\mathrm{p}} \pm \Delta m''_{\mathrm{p}}$. As mentioned above, the balance reliability $R_{T''}$ of the rotating surface T'' can be obtained. Therefore, the reliability of rotor dynamic balance is given as follows:

$$R_2 = \min\left(\left\{R_{T'},\ R_{T''}\right\}\right) \tag{21}$$

Example:

Unbalanced mass of a disc-shaped rotating part in machine tool exist. m_1=1kg±0.01kg, a_1=0°±0.01°, $r_1 = 100^{0.2}_{0}$ mm。 Added balance mass m_{p}=1kg±0.02kg, a_{p}=180°±0.2°, $r_p = 100^{0.1}_{0}$。 mm。 speed n=1 000r/min ±10 r/min。

$$\mu_r = 0.025,\ \ \sigma^2_r = 0.1436$$

$$\mu_{[r]} = 0.060\ 2,\ \ \sigma^2_{[r]} = 4.026 \times 10^{-8}$$

$$R_1 = \varPhi = (0.092\ 86) \approx 53.7\%$$

MONTE CARLO SIMULATION OF GEAR RELIABILITY CALCULATION

Based on the sensitivity analysis, the reliability calculation of contact fatigue strength and bending fatigue strength of gears is studied by Monte Carlo simulation.

Sensitivity analysis of gear contact stress:

The formula of gear contact stress is given as follows:

$$\sigma_H = Z_E \cdot Z_H \cdot Z_\varepsilon \cdot \sqrt{\frac{2KT_1}{bd_1^2} \cdot \frac{u+1}{u}} \tag{22}$$

where, T_1 is the torque transmitted by the pinion, d_1 is the diameter of the graduation circle of the pinion, u is the ratio of teeth, b is the tooth width, Z_E is the elastic coefficient, Z_H is the node area coefficient, Z_ε is the coincidence coefficient, and K is the load coefficient.

When considering the influence of random factors, consider,,, $T_1, d_1, K, b, Z_E, Z_H, Z_\varepsilon$ as random variables, their mean values are $\overline{T}_1, \overline{d}_1, \overline{K}, \overline{b}, \overline{Z}_E, \overline{Z}_H, \overline{Z}_\varepsilon$,,, and the mean value of gear contact stress is given as follows:

$$\overline{\sigma}_H = \overline{Z}_E \cdot \overline{Z}_H \cdot \overline{Z}_\varepsilon \cdot \sqrt{\frac{2\overline{K}\,\overline{T}_1}{\overline{b}\,\overline{d}_1^2} \cdot \frac{u+1}{u}} \tag{23}$$

The sensitivity of the mean value of gear contact stress to $\overline{T}_1, \overline{d}_1, \overline{K}, \overline{b}, \overline{Z}_E, \overline{Z}_H, \overline{Z}_\varepsilon$ is respectively

$$\frac{\partial \overline{\sigma}_H}{\partial \overline{Z}_E} = \overline{Z}_H \cdot \overline{Z}_\varepsilon \cdot \sqrt{\frac{2\overline{K}\,\overline{T}_1}{\overline{b}\,\overline{d}_1^2} \cdot \frac{u+1}{u}} \tag{24}$$

$$\frac{\partial \overline{\sigma}_H}{\partial \overline{Z}_H} = \overline{Z}_E \cdot \overline{Z}_\varepsilon \cdot \sqrt{\frac{2\overline{K}\,\overline{T}_1}{\overline{b}\,\overline{d}_1^2} \cdot \frac{u+1}{u}} \tag{25}$$

$$\frac{\partial \bar{\sigma}_H}{\partial \bar{Z}_\varepsilon} = \bar{Z}_E \cdot \bar{Z}_H \cdot \sqrt{\frac{2\bar{K}\bar{T}_1}{\overline{bd}_1^2} \cdot \frac{u+1}{u}} \tag{26}$$

$$\frac{\partial \bar{\sigma}_H}{\partial \bar{K}} = \bar{Z}_E \cdot \bar{Z}_H \cdot \bar{Z}_\varepsilon \cdot \sqrt{\frac{\bar{T}_1}{2\bar{K}\overline{bd}_1^2} \cdot \frac{u+1}{u}} \tag{27}$$

$$\frac{\partial \bar{\sigma}_H}{\partial \bar{T}_1} = \bar{Z}_E \cdot \bar{Z}_H \cdot \bar{Z}_\varepsilon \cdot \sqrt{\frac{\bar{K}}{2\bar{T}_1\overline{bd}_1^2} \cdot \frac{u+1}{u}} \tag{28}$$

$$\frac{\partial \bar{\sigma}_H}{\partial \bar{b}} = -\frac{\bar{Z}_E \bar{Z}_H \bar{Z}_\varepsilon}{2\bar{b}} \sqrt{\frac{2\bar{K}\bar{T}_1}{\overline{bd}_1^2} \cdot \frac{u+1}{u}} \tag{29}$$

$$\frac{\partial \bar{\sigma}_H}{\partial \bar{d}_1} = -\frac{\bar{Z}_E \bar{Z}_H \bar{Z}_\varepsilon}{\bar{d}_1^2} \sqrt{\frac{2\bar{K}\bar{T}_1}{\bar{b}} \cdot \frac{u+1}{u}} \tag{30}$$

Sensitivity analysis of gear bending stress:

The formula of gear bending stress is given as follows:

$$\sigma_F = \frac{2KT_1}{bd_1 m} Y_{Fa} \cdot Y_{Sa} \cdot Y_\varepsilon \tag{31}$$

where, m is the modulus, Y_{Fa} is the tooth shape coefficient, Y_{Sa} is the root stress correction coefficient, and Y_ε is the coincidence coefficient. When considering the influence of random factors, consider $T_1, d_1, K, b, Y_{Fa}, Y_{Sa}, Y_\varepsilon, m$ as random variables, their mean value is $\bar{T}_1, \bar{d}_1, \bar{K}, \bar{b}, \bar{Y}_{Fa}, \bar{Y}_{Sa}, \bar{Y}_\varepsilon, \bar{m}$ and the mean value of gear bending stress is given as follows:

$$\bar{\sigma}_F = \frac{2\bar{K}\bar{T}_1}{\overline{bd}_1\bar{m}} \bar{Y}_{Fa} \cdot \bar{Y}_{Sa} \cdot \bar{Y}_\varepsilon \tag{32}$$

The sensitivity of the mean value of gear bending stress to $\bar{K}, \bar{T}_1, \bar{Y}_{Fa}, \bar{Y}_{Sa}, \bar{Y}_\varepsilon, \bar{b}, \bar{d}_1, \bar{m}$ is respectively

$$\frac{\partial \bar{\sigma}_F}{\partial \bar{K}} = \frac{2\bar{T}_1}{b \bar{d}_1 \bar{m}} \bar{Y}_{Fa} \cdot \bar{Y}_{Sa} \cdot \bar{Y}_\varepsilon \tag{33}$$

$$\frac{\partial \bar{\sigma}_F}{\partial \bar{T}_1} = \frac{2\bar{K}}{b \bar{d}_1 \bar{m}} \bar{Y}_{Fa} \cdot \bar{Y}_{Sa} \cdot \bar{Y}_\varepsilon \tag{34}$$

$$\frac{\partial \bar{\sigma}_F}{\partial \bar{Y}_{Fa}} = \frac{2\bar{K}\bar{T}_1}{b \bar{d}_1 \bar{m}} \bar{Y}_{Sa} \cdot \bar{Y}_\varepsilon \tag{35}$$

$$\frac{\partial \bar{\sigma}_F}{\partial \bar{Y}_{Sa}} = \frac{2\bar{K}\bar{T}_1}{b \bar{d}_1 \bar{m}} \bar{Y}_{Fa} \cdot \bar{Y}_\varepsilon \tag{36}$$

$$\frac{\partial \bar{\sigma}_F}{\partial \bar{Y}_\varepsilon} = \frac{2\bar{K}\bar{T}_1}{b \bar{d}_1 \bar{m}} \bar{Y}_{Fa} \cdot \bar{Y}_{Sa} \tag{37}$$

$$\frac{\partial \bar{\sigma}_F}{\partial \bar{b}} = -\frac{2\bar{K}\bar{T}_1}{b^2 \bar{d}_1 \bar{m}} \bar{Y}_{Fa} \cdot \bar{Y}_{Sa} \cdot \bar{Y}_\varepsilon \tag{38}$$

$$\frac{\partial \bar{\sigma}_F}{\partial \bar{d}_1} = -\frac{2\bar{K}\bar{T}_1}{b \bar{d}_1^2 \bar{m}} \bar{Y}_{Fa} \cdot \bar{Y}_{Sa} \cdot \bar{Y}_\varepsilon \tag{39}$$

$$\frac{\partial \bar{\sigma}_F}{\partial \bar{m}} = -\frac{2\bar{K}\bar{T}_1}{b \bar{d}_1 \bar{m}^2} \bar{Y}_{Fa} \cdot \bar{Y}_{Sa} \cdot \bar{Y}_\varepsilon \tag{40}$$

Computer simulation of random variables:

If 12 uniformly distributed random numbers are generated, they are added up and then subtracted by 6, the sample values of standard normal variables can be obtained.

If $X_i \sim N\left(\mu_i, \sigma_i^2\right)$, $Z \sim N\left(0,1\right)$, using formulas

$$X_i = \mu_i + \sigma_i Z \tag{41}$$

Then we get the normal random variables X_i。

The random variable is subject to other distribution, and the sample value of the random variable can be obtained by the inverse transformation.

Calculation of reliability by Monte Carlo simulation method:

The application of Monte Carlo simulation in the calculation of mechanical reliability is to randomly select a group of values from the distribution of random variables, substitute them into the stress calculation formula to get a stress value, and then compare it with a strength value from the strength distribution. If the stress is greater than the strength, the part will fail; otherwise, the part will be safe. Set the simulation number as N, failure number as F, and reliability R=1-F/N. In order to obtain reliable simulation results, the greater the simulation times, the higher the simulation accuracy. As far as this paper is concerned, stress and strength refer to gear contact stress, gear contact fatigue strength or gear bending stress and gear bending fatigue strength.

Example:

The parameters of a gear drive are: $m = 2.5$, $z_1 = 23$, $z_2 = 106$, $\overline{T}_1 = 54N \cdot m$,

$\sigma_{T_1} = 0.4$, $\overline{Z}_H = 2.5$, $\sigma_{Z_H} = 0.02$, $\overline{Z}_E = 189.8\left(N/mm^2\right)^{\frac{1}{2}}$, $\sigma_{Z_E} = 1$, $\overline{Z}_\varepsilon = 0.88$,

$\sigma_{Z_\varepsilon} = 0.01$, $\overline{b} = 60$, $\sigma_b = 0.1$, $\overline{K} = 1.3$, $\sigma_K = 0.02$, $\overline{d}_1 = 57.5$, $\sigma_{d_1} = 0.05$,

$\left[\overline{S}_{H_1}\right] = 420MPa$, $\sigma_{\left[\overline{S}_{[H_1]}\right]} = 2$. Through sensitivity analysis, it is found that the

sensitivity of contact stress to T_1 is 3.084×10^{-8}, which is very small, so it is regarded as a constant. A computer program has been written with VB6.0 and simulated 1000 times. The reliability of contact fatigue strength of gears is 99.6%. The computer program is as follows:

Private Sub Command1_Click ()

Dim a(12) As Single

Dim N%, R!

```
Dim b(1000) As Single

Dim x1(1000) As Single

Dim x2(1000) As Single

Dim x3(1000) As Single

Dim x4(1000) As Single

Dim x5(1000) As Single

Dim x6(1000) As Single

Dim x7(1000) As Single

Dim x8(1000) As Single

Dim x9(1000) As Single

N = 0

For i = 1 To 1000

a(0) = 0

For j = 1 To 12

a(j) = a(j - 1) + Rnd

Next j

b(i) = a(12) - 6

x1(i) = 189.8 + 1 * b(i)

x2(i) = 2.5 + 0.02 * b(i)

x3(i) = 0.88 + 0.01 * b(i)
```

x4(i) = 1.3 + 0.02 * b(i)

x5(i) = 54000 + 400 * b(i)

x6(i) = 60 + 0.1 * b(i)

x7(i) = 57.5 + 0.05 * b(i)

x8(i) = 420 + 2 * b(i)

x9(i) = x1(i) * x2(i) * x3(i) * Sqr(2 * x4(i) * x5(i) * 5.609 / (x6(i) * x7(i) * x7(i) * 4.609))

If (x9(i) < x8(i)) Then

N = N + 1

End If

Next i

R = N / 1000

Print R

End Sub

RELIABILITY CALCULATION BASED ON SENSITIVITY ANALYSIS

Based on sensitivity analysis, this paper uses an optimization method to calculate reliability.

Optimization model:

If the limit state equation has n independent normal random variables

$X_1, X_2, \cdots, X_n,$ then the limit state equation is

$$Z = g(X_1, X_2, \cdots, X_n) = 0 \tag{42}$$

Make a standard normal transformation for normal variables $X_i (i = 1,2,\cdots,n)$, and

$$\hat{X}_i = \frac{X_i - \mu_{x_i}}{\sigma_{x_i}} \tag{43}$$

Therefore, in the standard normal coordinate system, the Surface equation in n-dimensional space is given as:

$$Z = g(\hat{X}_1 \sigma_{x_1} + \mu_{x_1}, \hat{X}_2 \sigma_{x_2} + \mu_{x_2}, \cdots, \hat{X}_n \sigma_{x_n} + \mu_{x_n}) = 0 \tag{44}$$

The shortest distance from the coordinate origin to the limit state surface is the safety index β value, then the reliability is $R = \Phi(\beta)$

The mathematical model of the index by optimization method

The design variable is given as follows:

$$\mathbf{X} = [\hat{X}_1, \hat{X}_2, \cdots, \hat{X}_n]^T \tag{45}$$

The objective function is given as follows:

$$\beta = \min f(\mathbf{X}) = \sqrt{\hat{X}_1^2 + \hat{X}_2^2 + \cdots + \hat{X}_n^2} \tag{46}$$

The constraints are:

$$g(\hat{X}_1 \sigma_{x_1} + \mu_{x_1}, \hat{X}_2 \sigma_{x_2} + \mu_{x_2}, \cdots, \hat{X}_n \sigma_{x_n} + \mu_{x_n}) = 0 \tag{47}$$

$$b_1 < \hat{X}_1 < c_1$$

$$b_2 < \hat{X}_2 < c_2$$

$$\vdots \quad \vdots \quad \vdots \tag{48}$$

$$b_n < \hat{X}_n < c_2$$

b_i , c_i are the lower and upper bounds of the range of $\hat{X}_i$, respectively

Sensitivity analysis:

The stress is expressed as

$$y = z(X_1, X_2, \cdots X_n) \tag{49}$$

Sensitivity of stress to design parameters (random variables X_i)

$$\frac{\partial y}{\partial X_i} = \frac{\partial z(X_1, X_2, \cdots, X_n)}{\partial X_i} \tag{50}$$
$$i = 1, 2, \cdots, n.$$

Sensitivity of stress to mean value of design parameters X_i

$$\frac{\partial y}{\partial X_i}\bigg|_{\bar{X}_i} = Z_1\left(\bar{X}_1, \bar{X}_2, \cdots, \bar{X}_n\right) \tag{51}$$

where, $i = 1, 2, \cdots, n.$

Through sensitivity analysis, we can find out the design parameters with high sensitivity and change these parameters when redesigning. The designer must give reasonable tolerance. When machining is needed, we should adopt the finishing method to assure the accuracy of design parameters and strengthen quality management and control. The design parameters with little sensitivity are found out, which can be treated as constants during the redesign, reducing the calculation amount and adopting a rough machining method when it is needed. The reliability calculation without sensitivity analysis is blind.

Optimization model based on sensitivity analysis:

Through the sensitivity analysis, the sensitivity of a design variable is very small, so it should be treated as a constant. Design parameters with high sensitivity should be modified to give new tolerances. The design parameters with general sensitivity can be modified or not according to the actual situation. The optimization model based on sensitivity analysis is;

Where, design variables are given as follows:

$$\mathbf{X} = [\hat{X}_1, \hat{X}_2, \cdots, \hat{X}_m]^T \tag{52}$$

The objective function is given as follows:

$$\beta = \min f_1(\mathbf{X}) = \sqrt{\hat{X}_1^2 + \hat{X}_2^2 + \cdots + \hat{X}_m^2} \tag{53}$$

The constraints are:

$$g_1(\hat{X}_1 \sigma_{x_1} + \mu_{x_1}, \hat{X}_2 \sigma_{x_2} + \mu_{x_2}, \cdots, \hat{X}_m \sigma_{x_m} + \mu_{x_m}) = 0 \tag{54}$$

$$\begin{aligned} a_1 &< \hat{X}_1 < e_1 \\ a_2 &< \hat{X}_2 < e_2 \\ \vdots \quad &\quad \vdots \quad \quad \vdots \\ a_m &< \hat{X}_m < e_m \end{aligned} \tag{55}$$

Multiple failure modes:

It is very common that many failure modes coexist in mechanical parts. In a variety of failure modes, because each failure mode is likely to occur, we consider that sensitivity analysis is carried out for each failure mode, and respective optimization models are established to calculate their reliability. If the sensitivity of a certain design parameter is different in different failure modes, it is suggested that the design parameter should be modified according to the greatest sensitivity of absolute value in the redesign.

The reliability calculation of multiple failure modes can also be solved by the multi-objective optimization method. The optimization model is as follows:

The design variables are:

$$\mathbf{X} = [\hat{X}_1, \hat{X}_2, \cdots, \hat{X}_j]^T \tag{56}$$

The objective functions are:

$$\beta_1 = \min f_1(\mathbf{X}) = \sqrt{\hat{X}_1^2 + \hat{X}_2^2 + \cdots + \hat{X}_{m_1}^2} \quad (m_1 \leq j)$$

$$\beta_2 = \min f_2(\mathbf{X}) = \sqrt{\hat{X}_1^2 + \hat{X}_2^2 + \cdots + \hat{X}_{m_2}^2} \quad (m_2 \leq j)$$

$$\vdots \qquad \vdots \qquad\qquad \vdots \qquad\qquad \vdots$$

$$\beta_q = \min f_q(\mathbf{X}) = \sqrt{\hat{X}_1^2 + \hat{X}_2^2 + \cdots + \hat{X}_{m_q}^2} \quad (m_q \leq j) \tag{57}$$

The constraints are:

$$g_1(\hat{X}_1\sigma_{x_1} + \mu_{x_1}, \hat{X}_2\sigma_{x_2} + \mu_{x_2}, \cdots, \hat{X}_{m_1}\sigma_{x_{m_1}} + \mu_{x_{m_1}}) = 0$$

$$g_2(\hat{X}_1\sigma_{x_1} + \mu_{x_1}, \hat{X}_2\sigma_{x_2} + \mu_{x_2}, \cdots, \hat{X}_{m_2}\sigma_{x_{m_2}} + \mu_{x_{m_2}}) = 0$$

$$\vdots \qquad\qquad\qquad \vdots$$

$$g_q(\hat{X}_1\sigma_{x_1} + \mu_{x_1}, \hat{X}_2\sigma_{x_2} + \mu_{x_2}, \cdots, \hat{X}_{m_q}\sigma_{x_{m_q}} + \mu_{x_{m_q}}) = 0 \tag{58}$$

$$s_1 < \hat{X}_1 < t_1$$

$$s_2 < \hat{X}_2 < t_2$$

$$\vdots \qquad \vdots \qquad \vdots \tag{59}$$

$$s_j < \hat{X}_j < t_j$$

Optimization algorithm:

The above single-objective optimization problems contain equality constraints and inequality constraints. The mixed penalty function method or augmented multiplier method is used to transform the constrained optimization problem into an unconstrained optimization problem. Conjugate gradient method, variable scale method and Powell method can be used for unconstrained optimization problems. The genetic algorithm can also be used in unconstrained optimization.

Example:

The parameters of spiral compression spring are:

$$\overline{d} = 2.0mm, \quad \sigma_d = 0.008, \quad \overline{D} = 16.0mm,$$

$$\sigma_D = 0.0928, \quad \overline{n} = 14, \sigma_n = 0.0833, \quad \overline{y} = 20mm,$$

$$\sigma_y = 0.4, \quad \overline{G} = 79250MPa, \quad \sigma_G = 1585, \text{ shear fatigue strength}$$

$$\overline{S}_{SN} = 375MPa, \quad \sigma_{S_{SN}} = 35.1_{\circ}$$

$$\overline{\tau} = \frac{(1+\dfrac{\overline{d}}{2\overline{D}})\overline{d}\ \overline{G}\ \overline{y}}{\pi \overline{D}^2 \overline{n}}$$

$$\frac{\partial \overline{\tau}}{\partial \overline{d}} = \frac{\overline{G}\ \overline{y}}{\pi\ \overline{n}}\left(\frac{1}{\overline{D}^2} + \frac{\overline{d}}{\overline{D}^3}\right) = 158.5$$

$$\frac{\partial \overline{\tau}}{\partial \overline{D}} = \frac{\overline{G}\ \overline{y}}{\pi\ \overline{n}}\left((-2)\frac{\overline{d}}{\overline{D}^3} + (-3)\frac{\overline{d}^2}{2\overline{D}^4}\right) = -38.5 \qquad \frac{\partial \overline{\tau}}{\partial \overline{G}} = \frac{\overline{y}}{\pi\ \overline{n}}\left(\frac{\overline{d}}{\overline{D}^2} + \frac{\overline{d}^2}{2\overline{D}^3}\right) = 0.00377$$

$$\frac{\partial \overline{\tau}}{\partial \overline{y}} = \frac{\overline{G}}{\pi\ \overline{n}}\left(\frac{\overline{d}}{\overline{D}^2} + \frac{\overline{d}^2}{2\overline{D}^3}\right) = 15.0$$

$$\frac{\partial \overline{\tau}}{\partial \overline{n}} = (-1)\frac{\overline{G}\ \overline{y}}{\pi\ \overline{n}^2}\left(\frac{\overline{d}}{\overline{D}^2} + \frac{\overline{d}^2}{2\overline{D}^3}\right) = -21.4$$

The sensitivity analysis shows that G can be regarded as a constant and D needs to be modified. The optimization model has σ_d modified to 0.004.

The design variables are:

$$\mathbf{X} = [\hat{d}, \hat{D}, \hat{y}, \hat{n}, \hat{S}_{SN}]^T$$

The objective function is:

$$\beta = \min f(\mathbf{X}) = \sqrt{\hat{d}^2 + \hat{D}^2 + \hat{y}^2 + \hat{n}^2 + \hat{S}_{SN}^2}$$

The constraints are:

$$35.1\hat{S}_{SN} + 375 - \frac{79250\left(1+\dfrac{2+0.004\hat{d}}{2(16+0.0928\hat{D})}\right)(2+0.004\hat{d})(20+0.4\hat{y})}{\pi(16+0.0928\hat{D})^2(14+0.0833\hat{n})} = 0$$

$$-3 < \hat{d} < 3$$
$$-3 < \hat{D} < 3$$
$$-3 < \hat{n} < 3$$
$$-3 < \hat{y} < 3$$
$$-3 < \hat{S}_{SN} < 3$$

The mixed penalty function method is used to transform the constrained optimization problem into an unconstrained optimization problem. The Powell method is used in unconstrained optimization.

$$\beta = 2.16 \ , \ R = \Phi(\beta) = 0.9846$$

THE RELIABILITY OF SPRING

The calculation formula of spring composite stress is

$$\tau = K\frac{8FC}{\pi d^2} \tag{60}$$

K is the curvature coefficient of spring, C is the winding ratio, or spring index. F is the axial force, d is the diameter of the spring wire.

$[\tau]$ is the allowable stress, or strength

The limit state equation is

$$g(X) = [\hat{\tau}]\sigma_{[\hat{\tau}]} + \mu_{[\hat{\tau}]} - \left(\hat{K}\sigma_{\hat{K}} + \mu_{\hat{K}}\right)\frac{8\left(\hat{F}\sigma_{\hat{F}} + \mu_{\hat{F}}\right)\left(\hat{C}\sigma_{\hat{C}} + \mu_{\hat{C}}\right)}{\pi\left(\hat{d}\sigma_{\hat{d}} + \mu_{\hat{d}}\right)^2} = 0 \tag{61}$$

Spring material performance parameters, geometric dimensions, loads, allowable stresses, or strengths are looked at as normal random variables. The strength reliability of spring can be calculated by HL-RF method

The design variables are:

$$\mathbf{X} = [\hat{K}, \hat{F}, \hat{C}, \hat{d}, [\hat{\tau}]]^T \tag{62}$$

The objective function is given as follows:

$$\beta = \min f(X) = \sqrt{\hat{K}^2 + \hat{F}^2 + \hat{C}^2 + \hat{d}^2 + [\hat{\tau}]^2} \tag{63}$$

The constraints are:

$$g(X) = [\hat{\tau}]\sigma_{[\hat{\tau}]} + \mu_{[\hat{\tau}]} - \left(\hat{K}\sigma_{\hat{K}} + \mu_{\hat{K}}\right)\frac{8\left(\hat{F}\sigma_{\hat{F}} + \mu_{\hat{F}}\right)\left(\hat{C}\sigma_{\hat{C}} + \mu_{\hat{C}}\right)}{\pi\left(\hat{d}\sigma_{\hat{d}} + \mu_{\hat{d}}\right)^2} = 0 \tag{64}$$

A mathematical optimization model for reliability R is established. The following iterative formula can be used to solve the optimization problem

$$X_{k+1} = \frac{1}{\left|\nabla g(X_k)\right|^2}\left[\nabla g(X_k)X_k - g(X_k)\right]\nabla g(X_k)^T \tag{65}$$

Where, $\nabla g(X_k)$ is gradient

$$\nabla g(X_k)^T = \begin{pmatrix} -\sigma_{\hat{K}} \dfrac{8\left(\hat{F}\sigma_{\hat{F}}+\mu_{\hat{F}}\right)\left(\hat{C}\sigma_{\hat{C}}+\mu_{\hat{C}}\right)}{\pi\left(\hat{d}\sigma_{\hat{d}}+\mu_{\hat{d}}\right)^2} \\[2em] -8\sigma_{\hat{F}}\left(\hat{K}\sigma_{\hat{K}}+\mu_{\hat{K}}\right)\dfrac{\left(\hat{C}\sigma_{\hat{C}}+\mu_{\hat{C}}\right)}{\pi\left(\hat{d}\sigma_{\hat{d}}+\mu_{\hat{d}}\right)^2} \\[2em] -\sigma_{\hat{C}}\left(\hat{K}\sigma_{\hat{K}}+\mu_{\hat{K}}\right)\dfrac{8\left(\hat{F}\sigma_{\hat{F}}+\mu_{\hat{F}}\right)}{\pi\left(\hat{d}\sigma_{\hat{d}}+\mu_{\hat{d}}\right)^2} \\[2em] 16\sigma_{\hat{d}}\left(\hat{K}\sigma_{\hat{K}}+\mu_{\hat{K}}\right)\dfrac{\left(\hat{F}\sigma_{\hat{F}}+\mu_{\hat{F}}\right)\left(\hat{C}\sigma_{\hat{C}}+\mu_{\hat{C}}\right)}{\pi\left(\hat{d}\sigma_{\hat{d}}+\mu_{\hat{d}}\right)^3} \\[2em] \sigma_{[\tau]} \end{pmatrix} \tag{66}$$

The reliability R is

$$R = \Phi(\beta) \tag{.67}$$

According to the iterative formula, the computer program is compiled, and the partial derivative in the formula can be calculated by the accurate method. Finally, the computer program is run to get the reliability value.

CONCLUDING REMARKS

The reliability formulas of the static and dynamic balance of rigid rotors are given. The sensitivity analysis of the contact stress and bending stress of the gear makes the designer understand the sensitivity of random variables. The reliability calculation of contact fatigue strength and bending fatigue strength of gears is studied by Monte Carlo simulation. The optimization model of reliability calculation by optimization method is given. A multi-objective optimization model for reliability calculation of multiple failure modes is presented. The reliability of spring is calculated by the HL-RF method.

CONSENT FOR PUBLICATION

Not applicable.

CONFLICT OF INTEREST

The author declares no conflict of interest, financial or otherwise.

ACKNOWLEDGEMENTS

Declared none.

REFERENCES

[1]   R. Rackwitz, and B. Fiessler, "Structural reliability under combined load sequences", *Comput. Struc.,* vol. 9, pp. 489-494, 1978.
[http://dx.doi.org/10.1016/0045-7949(78)90046-9]

[2]   A.M. Hasofer, and N.C. Lind, "Exact and invariant second-moment code format", *J. Engrg. Mech. Div.,* vol. 100, pp. 357-366, 1974.

[3]   K. Breitung, "Asymptotic approximations for multinormal integrals", *J. Eng. Mech.,* vol. 110, pp. 357-366, 1984.
[http://dx.doi.org/10.1061/(ASCE)0733-9399(1984)110:3(357)]

[4]   M. Shinozuka, "Basic analysis of structural safety", *J. Eng. Mech.,* vol. 109, no. 3, pp. 721-740, 1983.

[5]   R.E. Melcher, "search-based importance samples", *Struct. Saf.,* no. 9, pp. 117-1286, 1990.
[http://dx.doi.org/10.1016/0167-4730(90)90003-8]

[6]   R. Deoliya, and T.K. Datta, "Reliability analysis of a microwave tower for fluctuating mean wind with directional effect", *Reliab. Eng. Syst. Saf.,* vol. 67, pp. 257-267, 2000.
[http://dx.doi.org/10.1016/S0951-8320(99)00053-8]

[7]   W.A. Gray, and R.E. Melchers, "Modifications to the 'directional simulation in the load space' approach to structural reliability analysis", *Probab. Eng. Mech.,* vol. 21, pp. 148-158, 2006.
[http://dx.doi.org/10.1016/j.probengmech.2005.08.002]

[8]   L.V. Utkin, and I. Kozine, "On new cautious structural reliability models in the framework of imprecise probabilities", *Struct. Saf.,* vol. 32, pp. 411-416, 2010.
[http://dx.doi.org/10.1016/j.strusafe.2010.08.004]

[9]   Arne Bang Huseby, "ErikVanem, BentNatvig, "Alternative environmental contours for structural reliability analysis", *Struct. Saf.,* vol. 54, pp. 32-45, 2015.
[http://dx.doi.org/10.1016/j.strusafe.2014.12.003]

[10] K. Pesinis, "Kong FahTee, "Statistical model and structural reliability analysis for onshore gas transmission pipelines", *Eng. Fail. Anal.,* vol. 82, pp. 1-15, 2017.
[http://dx.doi.org/10.1016/j.engfailanal.2017.08.008]

[11] Z. Kala, "Global sensitivity analysis of reliability of structural bridge system", *Eng. Struct.,* vol. 194, pp. 36-45, 2019.
[http://dx.doi.org/10.1016/j.engstruct.2019.05.045]

[12] Azad Yazdani, Mohammad-Sadegh Shahidzadeh,Tsuyoshi Takada, "Bayesian networks for disaggregation of structural reliability", *Structural Safety,* vol. 82, 2008.

CHAPTER 3

Reliability-Based Optimization Design

Abstract: Considering the influence of random factors, using the HL-RF method to calculate the reliability, the reliability optimization design of the gearbox is studied. The reliability optimization design of the gearbox is studied by using the importance sampling. This paper puts forward the method of reliability fuzzy optimum design, which is the deepening of reliability, optimum design, and fuzzy optimum design. Under the influence of fuzziness, a multi-objective reliability fuzzy optimization model of the gearbox is established. By using the multi-objective fuzzy optimization method, the reliability fuzzy optimization design of the gearbox is carried out.

Keywords: Gearbox, HL-RF method, Importance sampling, Multi-objective reliability fuzzy optimization, Reliability-based optimization design.

INTRODUCTION

The combination of reliability design and optimization design forms the reliability optimization design, which can not only predict the reliability of the product quantitatively but also make the design parameters of the product to get the optimization solution. Reliability optimization design does not consider the influence of fuzzy factors on products, while fuzzy optimization design considers the transition from no permitted intermediary, and the feasible area is expanded. The combination of fuzzy optimization design and reliability design forms the reliability fuzzy optimization design, which not only considers the impact of random factors on the product, but also considers the impact of fuzzy factors on the product. The optimization results are often better than the general optimization design, reliability optimization design and fuzzy optimization design, which is the deepening of reliability optimization design and fuzzy optimization design.

A method to solve the optimization of multidisciplinary mechanical systems subject to reliability-based constraints is presented [1]. Interactive stability-oriented reliability optimization of an offshore jacket and truss-column is discussed [2]. A heuristic procedure is proposed for solving a nonlinear knapsack class of reliability optimization problems [3]. The paper deals with two optimization techniques to solve mixed reliability-based optimization problem of truss structures [4]. A global

Wenhui Mo

optimization method focused on gear vibration reduction is proposed [5]. A reliability-based analysis and design optimization framework are presented that account for stochastic variations [6]. A multi-objective reliability optimization problem for system reliability is considered [7]. Reliability-based design optimization of elastoplastic mechanical structures is carried out [8]. An integrated algorithms system is proposed to carry out the reliability-based design optimization of the offshore towers [9]. An efficient hybrid approach based on a genetic algorithm for solving mixed integer nonlinear reliability optimization problems is proposed [10]. The weighted average simulation technique is one of the latest techniques solving structural reliability problems [11]. A methodology to perform the multi-objective reliability-based design optimization for a coupled acoustic-structural system is proposed [12]. A scenario optimization framework for reliability-based design is introduced [13].

The gearbox is an important mechanical device. The HL-RF method to compute reliability is feasible. The reliability-based optimization design of gearbox using HL-RF is proposed.

Optimization design based on the importance sampling (IS) for the gearbox is proposed. A multi-objective reliability-based fuzzy optimization model for the gearbox is established.

RELIABILITY-BASED OPTIMIZATION DESIGN USING HL-RF

(Fig. **1**) shows a gearbox. There are 12 gears and 4 axles in the gear\box. Design variables are:

$$x = (m_1,\ z_1,\ z_2,\ m_2,\ z_3,\ z_4,\ m_3,\ z_5,\ z_6,\ m_4,\ z_7,\ z_8,\ m_5,\ z_9,\ z_{10},\ m_6,\ z_{11}$$
$$,\ z_{12},\ b_1,\ b_2,\ b_3,\ b_4,\ d_1,\ l_1,\ d_2,\ l_2,\ d_3,\ l_3,\ d_4,\ l_4)^{\mathrm{T}}$$

The objective function is the sum of the masses of all gears and axles in the gearbox. It is defined as

$$f(x) = \frac{1}{4}\pi b_1(m_1^2 z_1^2 - d_0^2)\rho + \frac{1}{4}\pi b_1(m_1^2 z_2^2 - d_1^2)\rho$$

$$+ \frac{1}{4}\pi b_2(m_2^2 z_3^2 - d_1^2)\rho + \frac{1}{4}\pi b_2(m_2^2 z_4^2 - d_2^2)\rho$$

$$+\frac{1}{4}\pi b_2(m_3^2 z_5^2 - d_1^2)\rho + \frac{1}{4}\pi b_2(m_3^2 z_6^2 - d_2^2)\rho \qquad (1)$$

$$+\frac{1}{4}\pi b_2(m_4^2 z_7^2 - d_1^2)\rho + \frac{1}{4}\pi b_2(m_4^2 z_8^2 - d_2^2)\rho$$

$$+\frac{1}{4}\pi b_3(m_5^2 z_9^2 - d_2^2)\rho + \frac{1}{4}\pi b_3(m_5^2 z_{10}^2 - d_3^2)\rho$$

$$+\frac{1}{4}\pi b_4(m_6^2 z_{11}^2 - d_3^2)\rho + \frac{1}{4}\pi b_4(m_6^2 z_{12}^2 - d_4^2)\rho$$

$$+\frac{1}{4}\pi d_1^2 l_1\rho + \frac{1}{4}\pi d_2^2 l_2\rho + \frac{1}{4}\pi d_3^2 l_3\rho + \frac{1}{4}\pi d_4^2 l_4\rho$$

where, the ρ is the material density.

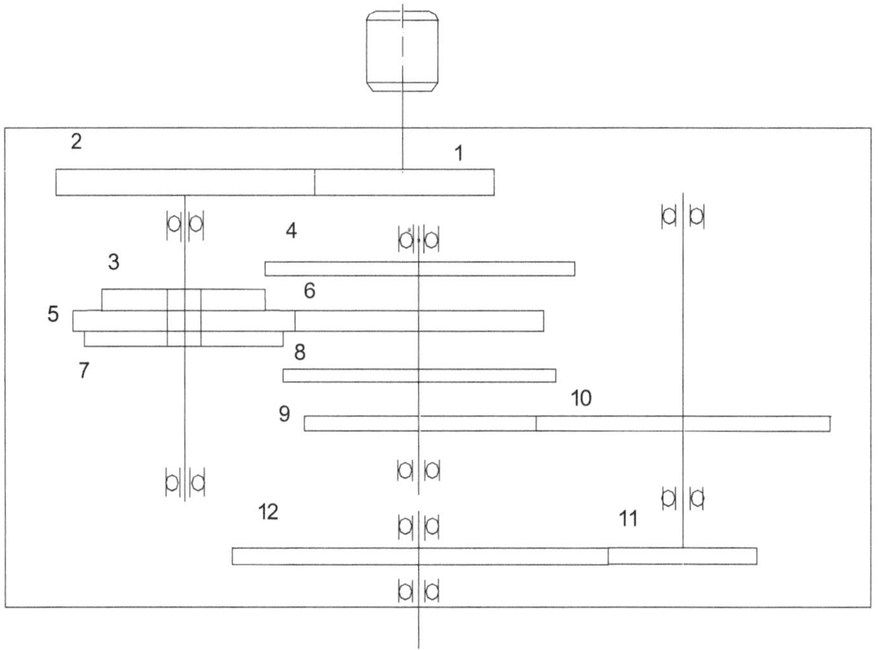

Fig. (1). A gearbox.

The constraints are defined as

$$U_{C_j} \geq 0.999 \ (j = 1, \ 3, \ 5, \ 7, \ 9, \ 11) \tag{2}$$

where, U_{C_j} is the reliability of contacting fatigue strength of small gear in each gear transmission.

$$U_{F_i} \geq 0.999 \ (i = 1, 2, \cdots, 12) \tag{3}$$

where, U_{F_i} is the reliability of bending fatigue strength of the gear.

$$U_{A_k} \geq 0.999 \qquad (k = \text{I, II, III, IV}) \tag{4}$$

where, U_{A_k} is the reliability of fatigue strength of the axle.

$$m_{kl} \leq m_k \leq m_{ks} \ (k = 1, \ 2, \ \ldots, \ 6) \tag{5}$$

$$z_{kl} \leq z_k \leq z_{ks} \ (k = 1, \ 2, \ \ldots, \ 12) \tag{6}$$

$$b_{kl} \leq b_k \leq b_{ks} \qquad (k = 1, \ 2, \ 3, \ 4) \tag{7}$$

$$d_{kl} \leq d_k \leq d_{ks} \qquad (k = 1, \ 2, \ 3, \ 4) \tag{8}$$

$$l_{kl} \leq l_k \leq l_{ks} \qquad (k = 1, \ 2, \ 3, \ 4) \tag{9}$$

where, $m_{kl}, z_{kl}, b_{kl}, d_{kl}, l_{kl}$ are the lower bounds. $m_{ks}, z_{ks}, b_{ks}, d_{ks}, l_{ks}$ are the upper bounds.

Reliability calculation:

Reliability index β can be formulated as a constrained optimization:

$$\text{Minimize } \beta = \sqrt{y^T y} \tag{10}$$

Subject to $G(y)=0$ **(11)**

The following recursive formula is generally used to solve the optimization problem [1].

$$y_{k+1} = \frac{1}{|\nabla G(y_k)|^2}\left[\nabla G(y_k)y_k - G(y_k)\right]\nabla G(y_k)^T$$ **(12)**

The reliability R is given by

$$R = \Phi(\beta)$$ **(13)**

where, $\Phi(\cdot)$ is the standard normal cumulative probability.

Using the penalty function method, a constrained optimization problem is transformed into an unconstrained optimization problem. The unconstrained optimization problem adopts the Powell method. Table **1** shows the comparison of design parameters.

Table 1. The comparison of design parameters.

	m_1	m_2	m_3	m_4	z_1
Original parameters	4	4	4	4	18
Optimal parameters	3	3.5	3.5	3.5	20
	z_2	z_3	z_4	z_5	z_6
Original parameters	44	27	43	35	35
Optimal parameters	37	19	36	25	30
	z_7	z_8	z_9	z_{10}	z_{11}
Original parameters	31	39	25	41	19
Optimal parameters	23	32	19	40	19

(Table 1) cont.....

	z_{12}	d_1	d_2	d_3	d_4
Original parameters	47	50	50	50	65
Optimal parameters	40	40	45	46	60
	m_5	m_6	b_1	b_2	b_3
Original parameters	4	4	25	25	25
Optimal parameters	4	4	18	20	25
	b_4	l_1	l_2	l_3	l_4
Original parameters	25	350	280	340	290
Optimal parameters	25	280	200	280	230

The original parameters refer to the original design parameters of the gearbox. Optimization parameters refer to the parameters of the gearbox obtained through the reliability-based optimization design using HL-RF. After the optimization design, the gearbox is lighter in weight, smaller in volume, more reasonable and more practical.

The Importance Sampling-Based Optimization Design for the Gearbox

(Fig. **1**) shows a gearbox. There are 12 gears and 4 axles in the gearbox.

Design variables and the objective function can be seen in section 3.2.

The constraints are defined as

$$P_{C_j} < 0.001 \ (j = 1, \ 3, \ 5, \ 7, \ 9, \ 11) \tag{14}$$

where, P_{C_j} is the probability of failure of contacting fatigue strength of small gear in each gear transmission.

$$P_{F_i} < 0.001 \ (i = 1, 2, \cdots, 12) \tag{15}$$

where, P_{F_i} is the probability of failure of bending fatigue strength of the gear.

$$P_{A_k} < 0.001 \qquad (k=\text{I,II,III,IV}) \tag{16}$$

where, P_{A_k} is the probability of failure of fatigue strength of the axle.

The lower bounds and the upper bounds of design variables can be seen in section 3.2.

Reliability of gear contact strength:

The formula of gear contact stress is as follows:

$$\sigma_H = Z_E \cdot Z_H \cdot Z_\varepsilon \cdot \sqrt{\frac{2KT_1}{bd_1^2} \cdot \frac{u+1}{u}} \tag{17}$$

Considering stochastic factors, $T_1, d_1, K, b, Z_E, Z_H, Z_\varepsilon$ are regarded as normal random variables.

The limit state function is given by:

$$g(x_1) = [\sigma_H] - Z_E \cdot Z_H \cdot Z_\varepsilon \cdot \sqrt{\frac{2KT_1}{bd_1^2} \cdot \frac{u+1}{u}} \tag{18}$$

The probability density function is given by:

$$f_X(x_1, x_2, \cdots, x_n) = \frac{1}{(2\pi)^{n/2}(\det C)} \exp\left\{ -\frac{1}{2}(X-\mu)' C^{-1}(X-\mu) \right\} \tag{19}$$

$$\text{where, } \mu = \begin{pmatrix} \mu_1 \\ \mu_2 \\ \vdots \\ \mu_n \end{pmatrix} = \begin{pmatrix} E(T_1) \\ E(d_1) \\ E(K) \\ E(b) \\ E(Z_E) \\ E(Z_H) \\ E(Z_\varepsilon) \end{pmatrix}, X = \begin{pmatrix} T_1 \\ d_1 \\ K \\ b \\ Z_E \\ Z_H \\ Z_\varepsilon \end{pmatrix}$$

E( ) is the symbol of solving the mean, C is the covariance matrix.

The probability of failure is

$$p_f = \int_{g(x)\leq 0} f_X(x)\, dx = \int_{g(x)\leq 0} f_X(x)\frac{\psi(x)}{\psi(x)}\, dx$$
$$\approx \frac{1}{N}\sum_{i=1}^{N} I\big(g(x_i)\leq 0\big)\frac{f_X(x_i)}{\psi(x_i)}$$

(20)

where $I\big(g(x_i)\leq 0\big)=1$ is the indicator function if x_i is in the failure domain and zero otherwise, $\psi(x_i)$ is the importance sampling probability density function.

The importance sampling probability density function is given by:

$$\psi(x_1, x_2, \cdots, x_n) = \ln\left\{-\frac{1}{2}(X-\mu)' C^{-1}(X-\mu)\right\}$$

(21)

Reliability of gear bending strength:

The formula of tooth root stress is as follows:

$$\sigma_F = \frac{2KT_1}{bd_1 m} Y_{Fa}\cdot Y_{Sa}\cdot Y_\varepsilon$$

(22)

Considering stochastic factors, $T_1, d_1, K, b, Y_{Fa}, Y_{Sa}, Y_\varepsilon, m$ are regarded as random variables.

The limit state function is given by:

$$g_1(x_2) = [\sigma_F] - \frac{2KT_1}{bd_1 m} Y_{Fa} \cdot Y_{Sa} \cdot Y_\varepsilon \tag{23}$$

The probabilistic density function is given by:

$$f_{X1}(x_1, x_2, \cdots, x_{n_1}) = \frac{1}{(2\pi)^{n_1/2}(\det C_1)} \exp\left\{-\frac{1}{2}(X_1 - \mu_1)' C_1^{-1}(X_1 - \mu_1)\right\} \tag{24}$$

$$\text{where, } \mu_{11} = \begin{pmatrix} E(T_1) \\ E(d_1) \\ E(K) \\ E(b) \\ E(Y_{Fa}) \\ E(Y_{Sa}) \\ E(Y_\varepsilon) \\ E(m) \end{pmatrix}, \ X_1 = \begin{pmatrix} T_1 \\ d_1 \\ K \\ b \\ Y_{Fa} \\ Y_{Sa} \\ Y_\varepsilon \\ m \end{pmatrix}$$

E () is the symbol of solving the mean, C is the covariance matrix.

The probability of failure is

$$p_{f_1} = \int_{g_1(x)\leq 0} f_{X_1}(x)\,dx = \int_{g_1(x)\leq 0} f_{X_1}(x)\frac{\psi_1(x)}{\psi_1(x)}\,dx$$
$$\approx \frac{1}{N_1}\sum_{i=1}^{N_1} I(g_1(x_i)\leq 0)\frac{f_X(x_i)}{\psi_1(x_i)} \tag{25}$$

where $I\left(g_1\left(x_i\right)\leq 0\right)=1$ is the indicator function if x_i is in the failure domain and zero otherwise, $\psi_1\left(x_i\right)$ is the importance sampling probability density function.

The importance sampling probability density function is given by

$$\psi_1\left(x_1,x_2,\cdots,x_{n_1}\right)=\ln\left\{-\frac{1}{2}\left(X_1-\mu_1\right)'C_1^{-1}\left(X_1-\mu_1\right)\right\} \tag{26}$$

Reliability of axle:

The formula of axle stress is as follows

$$\sigma_e=\frac{M_e}{W}=\frac{1}{0.1d^3}\sqrt{M^2+\left(\alpha T\right)^2} \tag{27}$$

where, σ_e is the axle stress, α is the conversion factor, W is the bending section modulus, M_e is the equivalent moment, M is the moment, T is the torque, d is the diameter of the axle.

The limit state function is given by:

$$g_2\left(x_3\right)=\left[\sigma_e\right]-\sigma_e \tag{28}$$

where, $\left[\sigma_e\right]$ is the allowable bending stress.

Considering stochastic factors, M, T, d are regarded as normal random variables.

The probabilistic density function is given by:

$$f_{11}\left(x_1,x_2,\cdots,x_{n_1}\right)=\frac{1}{\left(2\pi\right)^{n_1/2}\left(\det C_{11}\right)}\exp\left\{-\frac{1}{2}\left(X_{11}-\mu_{11}\right)'C_{11}^{-1}\left(X_{11}-\mu_{11}\right)\right\} \tag{29}$$

$$\text{where, } \mu_{11} = \begin{pmatrix} \mu_{11} \\ \mu_{21} \\ \mu_{31} \end{pmatrix} = \begin{pmatrix} E(M) \\ E(T) \\ E(d) \end{pmatrix}, \; X_{11} = \begin{pmatrix} M \\ T \\ d \end{pmatrix}$$

$E(\;)$ is the symbol of solving the mean, C_{11} is the covariance matrix.

The probability of failure is

$$p_{f_{11}} = \int_{g_{11}(x)\leq 0} f_{X_{11}}(x)\,dx = \int_{g_{11}(x)\leq 0} f_{X_{11}}(x)\frac{\psi_{11}(x)}{\psi_{11}(x)}\,dx$$
$$\approx \frac{1}{N_{11}}\sum_{i=1}^{N_{11}} I\big(g_{11}(x_i)\leq 0\big)\frac{f_{X_{11}}(x_i)}{\psi_{11}(x_i)} \tag{30}$$

where $I\big(g_{11}(x_i)\leq 0\big)=1$ is the indicator function if x_i is in the failure domain and zero otherwise, $\psi(x_i)$ is the importance sampling probability density function.

The importance sampling probability density function is given by:

$$\psi_{11}(x_1, x_2, \cdots, x_{n_{11}}) = \ln\left\{-\frac{1}{2}(X_{11}-\mu_{11})' C_{11}^{-1}(X_{11}-\mu_{11})\right\} \tag{31}$$

Using the penalty function method, a constrained optimization problem is transformed into unconstrained optimization problem. The unconstrained optimization problem adopts the Powell method. Table **2** shows the comparison of design parameters.

Table 2. The comparison of design parameters.

	m_1	m_2	m_3	m_4	z_1
Original parameters	4	4	4	4	18
Optimal parameters	3	3.5	3.5	3.5	20
	z_2	z_3	z_4	z_5	z_6

(Table 2) cont.....

Original parameters	44	27	43	35	35
Optimal parameters	37	19	36	25	30
	z_7	z_8	z_9	z_{10}	z_{11}
Original parameters	31	39	25	41	19
Optimal parameters	23	32	19	40	19
	z_{12}	d_1	d_2	d_3	d_4
Original parameters	47	50	50	50	65
Optimal parameters	40	40	45	46	60
	m_5	m_6	b_1	b_2	b_3
Original parameters	4	4	25	25	25
Optimal parameters	4	4	18	20	22
	b_4	l_1	l_2	l_3	l_4
Original parameters	25	350	280	340	290
Optimal parameters	25	280	200	260	200

The original parameters refer to the original design parameters of the gearbox. Optimization parameters refer to the parameters of the gearbox obtained through the importance sampling-based optimization. After the optimization design, the gearbox is lighter in weight, smaller in volume, more reasonable and more practical.

Multi-Objective Reliability-Based Fuzzy Optimization Design for the Gearbox

A multi-objective reliability-based fuzzy optimization model for the gearbox.

(Fig. **1**) shows a gearbox.

Design variables:

$$x=(m_1,\ z_1,\ z_2,\ m_2,\ z_3,\ z_4,\ m_3,\ z_5,\ z_6,\ m_4,\ z_7,\ z_8,\ m_5,\ z_9,\ z_{10},\ m_6,\ z_{11}$$
$$,\ z_{12},\ b_1,\ b_2,\ b_3,\ b_4,\ d_1,\ l_1,\ d_2,\ l_2,\ d_3,\ l_3,\ d_4,\ l_4)^{\mathrm{T}}$$

Objective functions:

The first objective function: the sum of the masses of all gears and shafts in the gearbox

$$f_1(x)=\frac{1}{4}\pi b_1(m_1^2 z_1^2 - d_0^2)\rho + \frac{1}{4}\pi b_1(m_1^2 z_2^2 - d_1^2)\rho$$

$$+\frac{1}{4}\pi b_2(m_2^2 z_3^2 - d_1^2)\rho + \frac{1}{4}\pi b_2(m_2^2 z_4^2 - d_2^2)\rho$$

$$+\frac{1}{4}\pi b_2(m_3^2 z_5^2 - d_1^2)\rho + \frac{1}{4}\pi b_2(m_3^2 z_6^2 - d_2^2)\rho$$

$$+\frac{1}{4}\pi b_2(m_4^2 z_7^2 - d_1^2)\rho + \frac{1}{4}\pi b_2(m_4^2 z_8^2 - d_2^2)\rho \qquad (32)$$

$$+\frac{1}{4}\pi b_3(m_5^2 z_9^2 - d_2^2)\rho + \frac{1}{4}\pi b_3(m_5^2 z_{10}^2 - d_3^2)\rho$$

$$+\frac{1}{4}\pi b_4(m_6^2 z_{11}^2 - d_3^2)\rho + \frac{1}{4}\pi b_4(m_6^2 z_{12}^2 - d_4^2)\rho$$

$$+\frac{1}{4}\pi d_1^2 l_1\rho + \frac{1}{4}\pi d_2^2 l_2\rho + \frac{1}{4}\pi d_3^2 l_3\rho + \frac{1}{4}\pi d_4^2 l_4\rho$$

where ρ is the material density, d_0=38mm.

The second objective function is the sum of the actual center distance of each gear transmission.

$$f_2(x) = \frac{m_1 z_1}{2} + \frac{m_1 z_2}{2} + \frac{m_2 z_3}{2} + \frac{m_2 z_4}{2} + \frac{m_5 z_9}{2} + \frac{m_5 z_{10}}{2} + \frac{m_6 z_{11}}{2} + \frac{m_6 z_{12}}{2} \tag{33}$$

The third objective function: the gear system is regarded as a series system (one pair of triple teeth is temporarily considered in the system and bending fatigue strength reliability of the other two pairs of gears are reflected in the constraint conditions). The system bending fatigue strength reliability (failure probability) is

$$f_3(x) = 1 - \Phi(U_{F1}) \cdot \Phi(U_{F2}) \cdot \Phi(U_{F3}) \cdot$$
$$\Phi(U_{F4}) \cdot \Phi(U_{F9}) \cdot \Phi(U_{F10}) \cdot \Phi(U_{F11}) \cdot \Phi(U_{F12}) \tag{34}$$

where, $\Phi(U_{Fi})$ is the value of U_{Fi}'s standard normal distribution function, $U_{Fi} = $

$$\frac{\ln(\overline{\sigma'}_{F\lim i} / \overline{\sigma'}_{Fi})^{[14]}}{\sqrt{C^2_{\sigma F\lim i} + C^2_{\sigma Fi}}} \quad (i=1,\ 2,\ 3,\ 4,\ 9,\ 10,\ 11,\ 12). \quad i \text{ represents gear number}$$

Constraint conditions:

The lower bound of the reliability of the contact fatigue strength of gears, the lower bound of the reliability of the infinite life fatigue strength of shafts, the lower bound of the reliability of the bending stiffness of output shafts and the upper and lower bound of each design variable are very difficult to give very accurate values. Because their values are affected by many fuzzy factors such as the design level, manufacturing level, material quality and service conditions, in fact, they should be considered the transition from no permitted intermediary. The ranges of upper and lower bounds are regarded as fuzzy subsets in design space.

The fuzzy constraints are as follows:

$$U_{pj} \geq \underset{\sim}{U}_{pc} \qquad (j = 1,\ 3,\ 5,\ 7,\ 9,\ 11) \tag{35}$$

$$U_{Rk_1} \geq \underset{\sim}{U}_R \qquad (k_1 = \text{I, II, III, IV}) \tag{36}$$

where j is the pinion number and k1 is the shaft number

$$\text{Us} \geq \underset{\sim}{U}_s \tag{37}$$

The above three formulas are respectively fuzzy constraint conditions of contact fatigue strength reliability of pinion in each pair of gears, fuzzy constraint conditions of infinite life fatigue strength reliability of each shaft and fuzzy constraint conditions of bending stiffness reliability of output shaft. Among them,

$$U_{pj} = \frac{\ln(\overline{\sigma}'_{H\lim j} / \overline{\sigma}_{Hj})^{[14]}}{\sqrt{c^2_{\sigma_{H\lim j}} + c^2_{\sigma_{Hj}}}} \tag{38}$$

$\underset{\sim}{U}_{pc}$ is the lower bound of U_{pj}

$$U_{Rk_1} = \frac{\overline{\sigma}'_{\lim k_1} - \overline{\sigma}_F{}^{[14]}}{\sqrt{S^2_{\sigma_{\lim k_1}} + S^2_{\sigma_F}}} \tag{39}$$

$\underset{\sim}{U}_R$ is the lower bound of U_{Rk_1}

$$U_S = \frac{\overline{y} - [y]^{[14]}}{S_y} \tag{40}$$

$\underset{\sim}{U}_s$ is the lower bound of U_s

The fuzzy constraints of each design variable are:

$$\underset{\sim}{m}_{kl} < m_k < \underset{\sim}{m}_{ku} \quad (k=1,\ 2,\ \ldots,\ 6) \tag{41}$$

$$\underset{\sim}{z}_{kl} < z_k < \underset{\sim}{z}_{ku} \quad (k=1,\ 2,\ \ldots,\ 12) \tag{42}$$

$$\underset{\sim}{b}_{kl} < b_k < \underset{\sim}{b}_{ku} \quad (k=1,\ 2,\ 3,\ 4) \tag{43}$$

$$\underset{\sim}{d}_{kl} < d_k < \underset{\sim}{d}_{ku} \quad (k=1,\ 2,\ 3,\ 4) \tag{44}$$

$$\underset{\sim}{l}_{kl} < l_k < \underset{\sim}{l}_{ku} \qquad (k=1,\ 2,\ 3,\ 4) \tag{45}$$

In order to simplify the calculation, the linear membership function is used. Using the horizontal cut set method, the fuzzy constraints can be transformed into general constraints as follows:

$$U_P \geq U_P^L + (U_P^U - U_P^L)\lambda^* \tag{46}$$

$$x \geq x^L + (x^U - \underline{x}^L)\lambda^* \tag{47}$$

$$x \leq \bar{x}U - (\bar{x}U - \bar{x}L)\lambda^* \tag{48}$$

where, x represents 30 independent design variables U_P^L, $\underline{U}_P^U$, $\bar{x}^U$, $\bar{x}^L$, $\underline{x}^U$, $\underline{x}^L$ are determined by the amplification coefficient method.

The third objective function only considers one pair of gears in the triple gear, and the other two pairs of gears are constrained by the reliability of bending fatigue strength

$$U_{F5} \geq [U_F]\ ;\ U_{F6} \geq [U_F]\ ;\ U_{F7} \geq [U_F]\ ;\ U_{F8} \geq [U_F]$$

where, $[U_F] = 3.72$, the corresponding reliability is 0.999 9。

Single objective fuzzy optimization reliability-based design:

In order to solve the problem of single objective fuzzy optimization with mixed constraints, it is necessary to transform the fuzzy constraints into common constraints by using the level cut set method, and then construct the constraints together with the original common constraints, so that the single objective fuzzy optimization problem can be transformed into a single objective common optimization problem.

The optimal values of the three objective functions f^*_j (j=1, 2, 3) of the above problems are obtained. Then, the membership function of the fuzzy optimal value is $M^j(f_j)=\left[\dfrac{\overline{f_j}-f_j}{\overline{f_j}-f^*_j}\right]^{\frac{1}{2}}$, where $\overline{f_i}$ is the maximum value f_i of the constraint. If f_i in $M^j(f_i)$ is replaced by $f_i(x)$, the membership function of the corresponding fuzzy optimal solution N_j is obtained:

$$N_j(x)=\left[\frac{\overline{f_j}-f_j(x)}{\overline{f_j}-f^*_j}\right]^{\frac{1}{2}} \tag{49}$$

The first objective function has a minimum value of 21.34 and a maximum value of 143.53. The second objective function has a minimum value of 367.5 and a maximum value of 900.4. The minimum value of the third objective function is very small, represented by 0, and the maximum value is 0.000031.

Fuzzy multi-objective optimization reliability-based design:

$$\text{let D}= \bigcap_{j=1}^{3} N_j$$

then the optimal solution x* of fuzzy optimization reliability-based for three targets should be made.

$$D(x^*)= \max_{x \in R}\left[\bigwedge_{j=1}^{3} N_j(x)\right] \tag{50}$$

In the formula, R is the feasible region of mixed constraint condition, which is difficult to be solved directly from (3.50). It can be transformed into the following single objective reliability fuzzy optimization problem

seek $\qquad x=(m_1,\ z_1,\ \ldots,\ l_4,\ a)^{\mathrm{T}}$

let max a

$$g_k(x) \leq 0$$

$$N_j(x) \geq a \qquad\qquad (51)$$

$$0 \leq a \leq 1$$

where a is an auxiliary variable and $g_k(x) \leq 0$ is a mixed constraint condition

$\lambda *$ determination and optimization results

$\lambda *$ is the optimal level value, which is determined by the two-level fuzzy comprehensive evaluation method. See Table **3** for influencing factors, factor grades and membership degrees.

Table 3. Influencing factors, factor grades and membership.

						Membership Degree				
	1	**2**	**3**	**4**	**5**	**1**	**2**	**3**	**4**	**5**
V_1	high	higher	common	lower	low	0.8	1.0	0.4	0.0	0.0
V_2	high	higher	common	lower	low	0.85	1.0	0.4	0.0	0.0
V_3	good	better	common	poor	bad	0.8	0.9	0.7	0.3	0.0
V_4	good	better	common	poor	bad	0.0	0.1	0.5	1.0	0.0
V_5	bad	poor	common	good	good	0.0	0.1	0.3	0.9	1.0
V_6	litter	less	common	bigger	big	0.0	0.3	0.7	0.5	0.2

Note: V_1 is the design level, V_2 is the manufacturing level, V_3 is the material quality, V_4 is the use condition, V_5 is the importance level, V_6 is the maintenance cost.

In general, fuzzy statistics or expert methods are used to determine the membership degree. An expert method is used. The alternative set is

$$\lambda = \{0.0, \ 0.1, \ 0.2, \ 0.3, \ 0.4, \ 0.5, \ 0.6, \ 0.7, \ 0.8, \ 0.9, \ 1.0\}$$

Since the rank order of each factor is arranged in accordance with the trend of influencing λ value, the rank evaluation matrix of each factor is

$$R_{\underset{\sim}{i}} = \begin{bmatrix} 0.2 & 0.25 & 0.3 & 0.35 & 0.4 & 0.45 & 0.5 & 0.8 & 1.0 & 0.5 & 0 \\ 0.1 & 0.15 & 0.2 & 0.25 & 0.3 & 0.4 & 0.45 & 0.9 & 1.0 & 0.8 & 0.1 \\ 0 & 0 & 0.1 & 0.3 & 0.4 & 0.5 & 0.6 & 0.8 & 0.9 & 1.0 & 0.2 \\ 0 & 0 & 0 & 0.1 & 0.3 & 0.5 & 0.1 & 0.85 & 0.95 & 0 & 0.5 \\ 0 & 0 & 0 & 0 & 0.1 & 0.3 & 0.5 & 0.7 & 1.0 & 1.0 & 0.8 \end{bmatrix}$$

where, $i=1, \ 2, \ 3, \ 4, \ 5, \ 6$.

The first level fuzzy comprehensive evaluation matrix is

$$R_{\underset{\sim}{}} = \begin{bmatrix} 0.118 & 0.159 & 0.218 & 0.295 & 0.355 & 0.436 & 0.495 & 0.846 & 0.982 & 0.728 & 0.082 \\ 0.12 & 0.161 & 0.22 & 0.297 & 0.356 & 0.437 & 0.496 & 0.844 & 0.982 & 0.722 & 0.08 \\ 0.093 & 0.124 & 0.181 & 0.276 & 0.355 & 0.451 & 0.531 & 0.838 & 0.968 & 0.784 & 0.141 \\ 0.0063 & 0.0095 & 0.044 & 0.172 & 0.331 & 0.494 & 0.653 & 0.838 & 0.938 & 0.987 & 0.381 \\ 0.0043 & 0.0065 & 0.022 & 0.089 & 0.226 & 0.409 & 0.589 & 0.78 & 0.967 & 0.991 & 0.574 \\ 0.0176 & 0.0264 & 0.076 & 0.197 & 0.318 & 0.459 & 0.591 & 0.821 & 0.944 & 0.965 & 0.341 \end{bmatrix}$$

The factor weight set is: $A_{\underset{\sim}{}} = (0.2, \ 0.22, \ 0.15, \ 0.10, \ 0.20, \ 0.13)$

Two level fuzzy comprehensive evaluation is $B_{\underset{\sim}{}} = A_{\underset{\sim}{}} \circ R_{\underset{\sim}{}}$

$$0.2, \ 0.22, \ 0.15, \ 0.10, \ 0.20, \ 0.13) \ R_{\underset{\sim}{}} = (0.068 \ \ 0.092 \ \ 0.138$$

$$0.226 \ \ 0.322 \ \ 0.44 \ \ 0.548 \ \ 0.827 \ \ 0.967 \ \ 0.844 \ \ 0.25)$$

According to the weighted average method, we obtain

$$\lambda^* = 0.666$$

Obviously, equation (3.50) and equation (3.51) are of the same solution, and they are still transformed into ordinary optimization problems by the horizontal cut set method. The penalty function method is used to solve the problem, and the comparison between the optimized result after rounding and the original design scheme is shown in Table **4**. The original parameters refer to the original design parameters of the gearbox. Optimization parameters refer to the parameters of the gearbox obtained through multi-objective reliability-based fuzzy optimization. After the optimization design, the gearbox is lighter in weight, smaller in volume, more reasonable and more practical.

Table 4. The comparison of design parameters.

	m_1	m_2	m_3	m_4	z_1
Original parameters	4	4	4	4	18
Optimal parameters	3	3.5	3.5	3.5	20
	z_2	z_3	z_4	z_5	z_6
Original parameters	44	27	43	35	35
Optimal parameters	37	19	36	25	30
	z_7	z_8	z_9	z_{10}	z_{11}
Original parameters	31	39	25	41	19
Optimal parameters	23	32	19	40	19
	z_{12}	d_1	d_2	d_3	d_4
Original parameters	47	50	50	50	65
Optimal parameters	40	40	45	46	60
	m_5	m_6	b_1	b_2	b_3
Original parameters	4	4	25	25	25
Optimal parameters	4	4	18	18	20

(Table 4) cont.....

	b_4	l_1	l_2	l_3	l_4
Original parameters	25	330	260	300	270
Optimal parameters	25	260	180	250	190

CONCLUDING REMARKS

The HL-RF method that computes the reliability and reliability optimization design of the gearbox is presented. The IS to compute reliability is feasible. Optimization design based IS of the gearbox is very efficient. A multi-objective reliability-based fuzzy optimization design of the gearbox is carried out, and the effect is very significant. It is a more scientific and practical design method to consider not only random factors but also the influence of fuzzy factors on mechanical products. The reliability fuzzy optimization design of the gearbox has a good effect. The product has light weight, compact structure, good rigidity, high reliability, and strong market competitiveness.

CONSENT FOR PUBLICATION

Not applicable.

CONFLICT OF INTEREST

The author declares no conflict of interest, financial or otherwise.

ACKNOWLEDGEMENTS

Declared none.

REFERENCES

[1]   D.R. Oakley, R.H. Sues, and G.S. Rhodes, "Performance optimization of multidisciplinary mechanical systems subject to uncertainties", *Probab. Eng. Mech.,* vol. 13, pp. 15-26, 1998. [http://dx.doi.org/10.1016/S0266-8920(97)00004-0]

[2]   M. Kleiber, A. Siemaszko, and R. Stocki, "Interactivestability-oriented reliability-based design optimization", *Comput. Methods Appl. Mech. Eng.,* vol. 168, pp. 243-253, 1999. [http://dx.doi.org/10.1016/S0045-7825(98)00143-1]

[3]   H. Ohtagaki, A. Iwasaki, Y. Nakagawa, and H. Narihisa, *"Smart greedy procedure for solving a multidimensional nonlinear knapsack class of reliability optimization problems"*, *Mathematical and Computer Modelling.*, vol. 31. December, 2000, pp. 283-288.

[4]   R. Stocki, K. Kolanek, S. Jendo, and M. Kleiber, "Study on discrete optimization techniques in reliability-based optimization of truss structures", *Comput. Struc.*, vol. 79, pp. 2235-2247. [http://dx.doi.org/10.1016/S0045-7949(01)00080-3]

[5]   M. Faggioni, *Farhad S.Samani,Gabriele Bertacchi,Francesco Pellicano, "Dynamic optimization of spur gears"*, *Mechanism and Machine Theory.*, vol. 46. December, 2011, pp. 544-557.

[6]   M. Allen, *Michael Raulli,Kurt Maute,Dan M.Frangopol, "Reliability-based analysis and design optimization of electrostatically actuated MEMS",Computers&Structures.*, vol. 82. December, 2004, pp. 1007-1020.

[7]   G.S. Mahapatra, and T.K. Roy, "Fuzzy multi-objective mathematical programming on reliability optimization model", *Appl. Math. Comput.*, vol. 174, pp. 643-659, 2006. [http://dx.doi.org/10.1016/j.amc.2005.04.105]

[8]   I. Kaymaz, and K. Marti, "Reliability-based design optimization for elastoplastic mechanical structures", *Comput. Struc.*, vol. 85, pp. 615-625, 2007. [http://dx.doi.org/10.1016/j.compstruc.2006.08.076]

[9]   H. Karadeniz, V. Toğan, and T. Vrouwenvelder, "An integrated reliability-based design optimization of offshore towers", *Reliab. Eng. Syst. Saf.*, vol. 94, pp. 1510-1516, 2009. [http://dx.doi.org/10.1016/j.ress.2009.02.008]

[10]   L. Sahoo, A. Banerjee, A.K. Bhunia, and S. Chattopadhyay, "An efficient GA–PSO approach for solving mixed-integer nonlinear programming problem in reliability optimization", *Swarm Evol. Comput.*, vol. 19, pp. 43-51, 2014. [http://dx.doi.org/10.1016/j.swevo.2014.07.002]

[11]   N.M. Okasha, "An improved weighted average simulation approach for solving reliability-based analysis and design optimization problems", *Struct. Saf.*, vol. 60, pp. 47-55, 2016. [http://dx.doi.org/10.1016/j.strusafe.2016.01.005]

[12]   Dammak Khalil, and Abdelkhalak El Hami, "Multi-objective reliability based design optimization of coupled acoustic-structural system", *Engineering Structures,* vol. 197, p. October, 2019.

[13]   Roberto Rocchetta, Luis G.Crespo, and Sean P.Kenny, "A scenario optimization approach to reliability-based design", *Reliability Engineering & System Safety,* vol. 196, p. March, 2020.

[14]   W. Mo, *"Reliability-fuzzy optimization design of dynamic box"*, *Mechanical Science and Technology,* vol. 17. December, 1996, pp. 957-958.

CHAPTER 4

Perturbation Stochastic Finite Element of Mechanical Vibration

Abstract: Perturbation stochastic finite element of mechanical vibration is introduced. The vibration equation of a system is transformed into a static problem by using the Willson θ method. An improved method of perturbation stochastic finite element is proposed. The formulas of calculating perturbation stochastic finite elements are transformed into linear equations to avoid solving matrix inversion. A numerical example is given in order to demonstrate the proposed.

Keywords: Linear equations, Matrix inversion, Newmark method, Perturbation stochastic finite element.

INTRODUCTION

The finite element method is a common method of calculating the vibration of the structure, but it neglects the influence of random factors. Material properties, geometry parameters and applied loads of structure are assumed to be stochastic. The stochastic finite element has been developed to deal with the engineering problem.

The direct Monte Carlo simulation takes a lot of computing time [1]. To improve the computational efficiency, the Neumann expansion is used and the NSFEM is proposed [2]. The preconditioned Conjugate Gradient method (PCG) is used to analyze a static problem [3]. The probability density function of the engineering problem is not easy to decide. The Perturbation stochastic finite element [4-10] is used by many authors with the perturbation technique. A hybrid spectral and metamodeling approach for the stochastic finite element analysis of dynamic structural systems is proposed [11]. Moreover, spectral stochastic finite element vibration analysis of fiber-reinforced composites is suggested [12]. A novel stochastic process to model the variation of rock strength is proposed [13]. In addition, robust characterization of the vibrational behaviour with random parameters is presented [14].

Wenhui Mo

Long CPU times for solving matrix inversion influence the computational efficiency of perturbation stochastic finite element. The perturbation stochastic finite element is improved to save computational time.

PERTURBATION STOCHASTIC FINITE ELEMENT OF MECHANICAL VIBRATION

Long CPU times for solving matrix inversion influence the computational efficiency of perturbation stochastic finite element. Therefore, the perturbation stochastic finite element is improved to save computational time.

Material properties, geometry parameters and applied loads of structures are regarded as normal random variables, and are indicated as $a_1, a_2, \cdots, a_i, \cdots, a_n$. Their means are $\mu_1, \mu_2, \cdots, \mu_i, \cdots, \mu_n$, and their variances are $\sigma_1^{2}, \sigma_2^{2}, \cdots, \sigma_i^{2}, \cdots, \sigma_n^{2}$.

For a linear system, the dynamic equilibrium equation is given by:

$$[M]\{\ddot{\delta}\} + [C]\{\dot{\delta}\} + [K]\{\delta\} = \{F\} \tag{1}$$

where $[M], [K]$ and $[C]$ are the global mass, stiffness and damping matrices.

At time $t + \Delta t$, the displacement vector using the Newmark method is given by

$$[\tilde{K}]\{\delta_{t+\Delta t}\} = \{\tilde{F}_{t+\Delta t}\} \tag{2}$$

$[\tilde{K}], \{\delta_{t+\Delta t}\}$ and $\{\tilde{F}_{t+\Delta t}\}$ are expanded about a_i *via* Taylor series

$$[\tilde{K}] = [\tilde{K}]^{0} + \sum_{i=1}^{n} \frac{\partial [\tilde{K}]}{\partial a_i} \alpha_i + \sum_{i=1}^{n} \frac{\partial^{2} [\tilde{K}]}{\partial a_i^{2}} \alpha_i^{2} + \cdots \tag{3}$$

$$\{\delta_{t+\Delta t}\} = \{\delta_{t+\Delta t}\}^{0} + \sum_{i=1}^{n} \frac{\partial \{\delta_{t+\Delta t}\}}{\partial a_i} \alpha_i + \sum_{i=1}^{n} \frac{\partial^{2} \{\delta_{t+\Delta t}\}}{\partial a_i^{2}} \alpha_i^{2} + \cdots \tag{4}$$

$$\left\{ F_{t+\Delta t} \right\} = \left\{ F_{t+\Delta t} \right\}^{0} + \sum_{i=1}^{n} \frac{\partial \left\{ F_{t+\Delta t} \right\}}{\partial a_{i}} \alpha_{i} + \sum_{i=1}^{n} \frac{\partial^{2} \left\{ F_{t+\Delta t} \right\}}{\partial a_{i}^{2}} \alpha_{i}^{2} + \cdots \tag{5}$$

By applying the perturbation technique, the following equations are given by:

$$\left\{ \delta_{t+\Delta t} \right\}^{0} = \left(\left[\tilde{K} \right]^{0} \right)^{-1} \left\{ F_{t+\Delta t} \right\}^{0} \tag{6}$$

$$\frac{\partial \left\{ \delta_{t+\Delta t} \right\}}{\partial a_{i}} = \left[\tilde{K} \right]^{-1} \left(\frac{\partial \left\{ \tilde{F}_{t+\Delta t} \right\}}{\partial a_{i}} - \frac{\partial \left[\tilde{K} \right]}{\partial a_{i}} \left\{ \delta_{t+\Delta t} \right\} \right) \tag{7}$$

$$\frac{\partial^{2} \left\{ \delta_{t+\Delta t} \right\}}{\partial a_{i} \partial a_{j}} = \left[\tilde{K} \right]^{-1} \left(\frac{\partial^{2} \left\{ \tilde{F}_{t+\Delta t} \right\}}{\partial a_{i} \partial a_{j}} - \frac{\partial \left[\tilde{K} \right]}{\partial a_{i}} \frac{\partial \left\{ \delta_{t+\Delta t} \right\}}{\partial a_{j}} \right.$$

$$\left. - \frac{\partial \left[\tilde{K} \right]}{\partial a_{j}} \frac{\partial \left\{ \delta_{t+\Delta t} \right\}}{\partial a_{i}} - \frac{\partial^{2} \left[\tilde{K} \right]}{\partial a_{i} \partial a_{j}} \left\{ \delta_{t+\Delta t} \right\} \right) \tag{8}$$

where $\dfrac{\partial \left\{ F_{t+\Delta t} \right\}}{\partial a_{i}}$ and $\dfrac{\partial^{2} \left\{ \tilde{F}_{t+\Delta t} \right\}}{\partial a_{i}^{2}}$ can be found in [7].

The mean of the displacement is obtained as:

$$\mu_{\delta} \approx \left\{ \delta_{t+\Delta t} \right\} \Big|_{a=\bar{a}} + \frac{1}{2} \sum_{i=1}^{n} \frac{\partial^{2} \left\{ \delta_{t+\Delta t} \right\}}{\partial a_{i}^{2}} \Bigg|_{a=\bar{a}} \cdot \sigma_{i}^{2} \tag{9}$$

The variance of $\left\{ \delta_{t+\Delta t} \right\}$ is given by:

$$Var_{\delta} \approx \sum_{i=1}^{n} \left(\frac{\partial \left\{ \delta_{t+\Delta t} \right\}}{\partial a_{i}} \Bigg|_{a=\bar{a}} \right)^{2} \cdot \sigma_{i}^{2} \tag{10}$$

$$\frac{\partial\{\ddot{\delta}_{t+\Delta t}\}}{\partial a_i}, \frac{\partial\{\dot{\delta}_{t+\Delta t}\}}{\partial a_i}, \frac{\partial^2\{\ddot{\delta}_{t+\Delta t}\}}{\partial a_i \partial a_j} \text{ and } \frac{\partial^2\{\dot{\delta}_{t+\Delta t}\}}{\partial a_i \partial a_j} \text{ can be found in Chapter 4, Ref. 7.}$$

The mean value and variance of the displacement are obtained at time $t+i_1\Delta t\left(i_1 = 2,3,\cdots,n_3\right)$ step-by-step.

AN IMPROVED METHOD OF PERTURBATION STOCHASTIC FINITE ELEMENT

It is well known that matrix inversions require long CPU times. $\left(\mathrm{K}_0\right)^{-1}$ is not easy to solve.

Eqs. 6, 7 and 8 are rewritten as:

$$\left[\tilde{K}\right]^0\{\delta_{t+\Delta t}\}^0 = \{F_{t+\Delta t}\}^0 \tag{11}$$

$$\left[\tilde{K}\right]\frac{\partial\{\delta_{t+\Delta t}\}}{\partial a_i} = \frac{\partial\{\tilde{F}_{t+\Delta t}\}}{\partial a_i} - \frac{\partial\left[\tilde{K}\right]}{\partial a_i}\{\delta_{t+\Delta t}\} \tag{12}$$

$$\left[\tilde{K}\right]\frac{\partial^2\{\delta_{t+\Delta t}\}}{\partial a_i \partial a_j} = \frac{\partial^2\{\tilde{F}_{t+\Delta t}\}}{\partial a_i \partial a_j} - \frac{\partial\left[\tilde{K}\right]}{\partial a_i}\frac{\partial\{\delta_{t+\Delta t}\}}{\partial a_j}$$

$$-\frac{\partial\left[\tilde{K}\right]}{\partial a_j}\frac{\partial\{\delta_{t+\Delta t}\}}{\partial a_i} - \frac{\partial^2\left[\tilde{K}\right]}{\partial a_i \partial a_j}\{\delta_{t+\Delta t}\} \tag{13}$$

Eqs.11, 12 and 13 are linear equations. The Cholesky decomposition is an effective method for solving linear equations. Using the Cholesky decomposition, $\{\delta_{t+\Delta t}\}^0$, $\frac{\partial\{\delta_{t+\Delta t}\}}{\partial a_i}$, $\frac{\partial^2\{\delta_{t+\Delta t}\}}{\partial a_i^2}$ can be obtained at the mean value point $\bar{a} = \left(\bar{a}_1,\bar{a}_2,\cdots,\bar{a}_i,\cdots,\bar{a}_n\right)^T$. Using Eqs. 9 and 10, the mean and variance of the displacement can be obtained.

Table 1. Parameters of the cantilever beam.

	Mean	Standard Deviation
Length	0.1m	0.000005
Width	0.01m	0.000004
Height	0.005m	0.000003
Poisson's ratio	0.2	0.01
Young's modulus	$2 \times 10^{11} \ N/m^2$	10^9
Load	100N	1

Table 2. CPU times of above-mentioned methods.

	CPU Time
DSFEM	48 minutes 17 seconds
PSFEM	2 hours 12 minutes 34 seconds
IPSFEM	30 minutes 46 seconds

Numerical example:

(Fig. **1**) shows a cantilever beam. The load subjected to the cantilever beam is Fsin(100t) N. It is divided into 40 rectangle elements that have 55 nodes. Table **1** shows the parameters of the cantilever beam. The length, width, height, the Poisson's ratio, Young's modulus and the load F are assumed to be normal random variables. (Fig. **1**) shows the mean of vertical displacement at node 5 within six seconds. (Fig. **2**) shows the variance of vertical displacement at node 55 within six seconds. The results produced by the PSFEM and the IPSFEM are close to those produced by the DSFEM. Table **2** shows the CPU times of the direct Monte Carlo simulation of the stochastic finite element method (DSFEM), second-order perturbation method of the stochastic finite element (PSFEM), improved method of perturbation stochastic finite element (IPSFEM) when the cantilever beam vibrated for six seconds. The IPSFEM requires the least amount of CPU time. The PSFEM requires the greatest amount of CPU time.

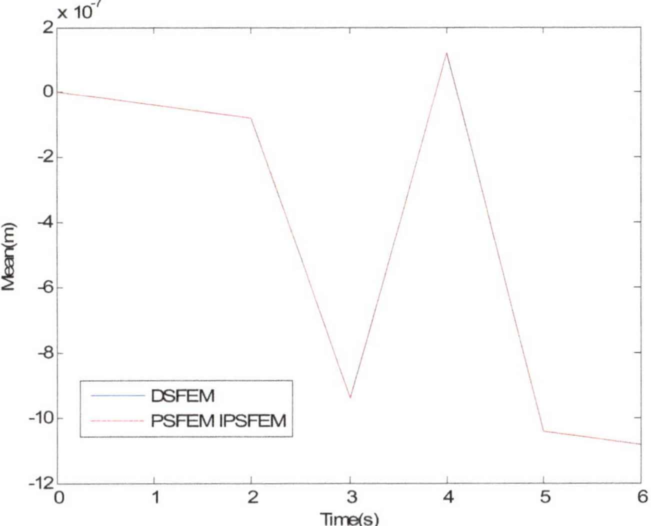

Fig. (1). The mean of vertical displacement at node 55.

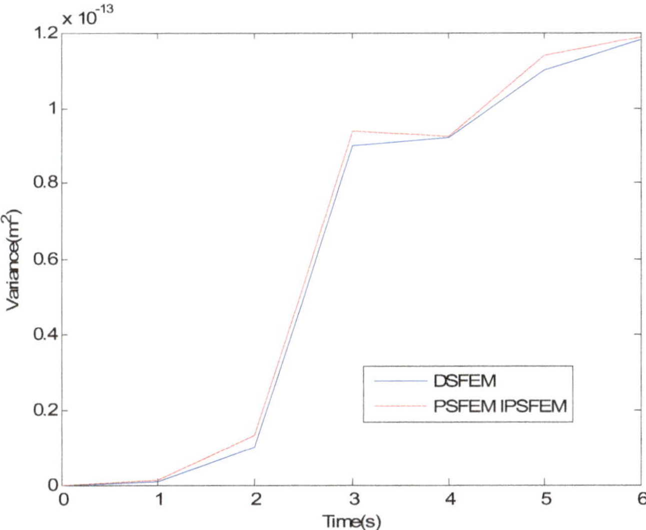

Fig. (2). The variance of vertical displacement at node 55.

CONCLUDING REMARKS

An improved method of perturbation stochastic finite element is proposed to save computational time. The calculation of perturbation stochastic finite element was carried out to transform into linear equations. The proposed method is applicable to stochastic structure problems. An example is given and calculated results are compared to confirm the proposed method.

CONSENT FOR PUBLICATION

Not applicable.

CONFLICT OF INTEREST

The author declares no conflict of interest, financial or otherwise.

ACKNOWLEDGEMENTS

Declared none.

REFERENCES

[1]　　J. Astill, C.J. Nosseir, and M. Shinozuka, "Impact loading on structures with random properties", *J. Struct. Mech.,* vol. 1, pp. 63-67, 1972.
[http://dx.doi.org/10.1080/03601217208905333]

[2]　　F. Yamazaki, M. Shinozuka, and G. Dasgupta, "Neumann expansion for stochastic finite element analysis", *ASCE J. Engng. Mech.,* vol. 114, pp. 1335-1354, 1988.
[http://dx.doi.org/10.1061/(ASCE)0733-9399(1988)114:8(1335)]

[3]　　M. Papadrakakis, and V. Papadopoulos, "Robust and efficient methods for stochastic finite element analysis using Monte Carlo Simulation", *Comput. Methods Appl. Mech. Eng.,* vol. 134, pp. 325-340, 1996.
[http://dx.doi.org/10.1016/0045-7825(95)00978-7]

[4]　　X.Q. Peng, Liu Geng, Wu Liyan, G.R. Liu, and K.Y. Lam, "A stochastic finite element method for fatigue reliability analysis of gear teeth subjected to bending", *Computational Mechanics,* pp. 253-261, 1998.

[5]　　W.K. Liu, T. Belytschko, and A. Mani, *"Random field finite element", J Nume Meth Engng.,* vol. 23. December, 1986, pp. 1831-1845.]

[6]　　W.K. Liu, T. Belytschko, and A. Mani, *"Probabilistic finite elements for nonlinear structural dynamics", Comput. Methods Appl. Mech. Engrg.,* vol. 57. January, 1986, pp. 61-81.

[7]　　W. Mo, "Stochastic finite element for structural vibration", *Math. Probl. Eng.,* vol. 2010, pp. 22-40, 2010.

[8]　　W. Mo, "Dynamic analysis of stochastic finite element based on sensitivity computation", *2010 International conference on Smart Materials and Intelligent Systems,* 2011pp. 1264-1269

[9]   Shashank Vadlamani, "Arun C.O. A, "Stochastic B-spline wavelet on the interval finite element method for problems in elasto-statics", *Probab. Eng. Mech.,* vol. 58, 2019.

[10]   B. Pokusiński, and M. Kamiński, "Lattice domes reliability by the perturbation-based approaches *vs.* semi-analytical method", *Comput. Struc.,* vol. 221, pp. 179-192, 2019.
[http://dx.doi.org/10.1016/j.compstruc.2019.05.012]

[11]   A. Kundu, F.A.Diaz Dela O.S.Adhikari, and M.I.Friswell., *"A hybrid spectral and met modeling approach for the stochastic finite element analysis of structural dynamic systems"*, Computer Methods in Applied Mechanics and Engineering., vol. 270. January, 2014, pp. 201-219.

[12]   K. Sepahvand, "Spectral stochastic finite element vibration analysis of fiber-reinforced composites with random fiber orientation", *Compos. Struct.,* vol. 145, pp. 119-128, 2016.
[http://dx.doi.org/10.1016/j.compstruct.2016.02.069]

[13]   D.M. Lobo, T.G. Ritto, and D.A. Castello, "A novel stochastic process to model the variation of rock strength in bit-rock interaction for the analysis of drill-string vibration", *Mech. Syst. Signal Process.,* vol. 141, 2019.

[14]   M. Ghienne, L. Laurent, and C. Blanzé, "Robust characterization of the vibrational behaviour of light assembled structures with random parameters", *Mech. Syst. Signal Process.,* vol. 136, 2020.
[http://dx.doi.org/10.1016/j.ymssp.2019.106510]

CHAPTER 5

Stochastic Finite Element Method of Dynamic Reliability

Abstract: Samples of material property are generated by the calculation of the stochastic field. Differential equations are transformed into linear equations by the Willson θ method. Linear equations are solved by the Successive Over Relaxation method. Dynamic reliability of structure can be given by Monte Carlo simulation. A new method of calculating dynamic reliability using the Neumann stochastic finite element is proposed.

Keywords: Dynamic reliability, Neumann stochastic finite element, Successive Over Relaxation method.

INTRODUCTION

Structural vibration is influenced by stochastic factors, such as material properties, geometry parameters, and applied loads. A tool to analyze structure with uncertainty is a stochastic finite element. Dynamic reliability should be considered in an engineering problem.

Structural vibration is analyzed by the direct Monte Carlo simulation of the stochastic finite element method (DSFEM) [1]. Dynamic analysis of structures often adopts the Perturbation stochastic finite element [2]. Monte Carlo simulation of the stochastic finite element method by applying the Neumann expansion (NSFEM) is introduced in the dynamic analysis [3]. The preconditions Conjugate Gradient method (PCG) is applied in dynamic analysis of structures [4, 5].

Stochastic finite element and the first-order second-moment method analyze dynamic reliability of frame [6]. The stochastic finite element calculates the reliability of gear teeth [7]. Importance sampling for finite element reliability analysis is suggested [8]. The reliability assessment of uncertain linear structures subjected to Gaussian distribution is proposed [9]. The main aim is to present the Stochastic perturbation-based Finite Element Method analysis of the stability and reliability of the underground steel vertical cylindrical structure of the waste container [11]. A scheme for integrating a model reduction technique into a

Wenhui Mo

simulation-based method for reliability sensitivity analysis of a class of medium/large nonlinear finite element models under stochastic excitation is presented [12]. This paper proposes a novel multi-scale approach for the reliability analysis of composite structures that accounts for both microscopic and macroscopic uncertainties, such as constituent material properties and ply angle [13]. This paper presents a practical approach for reliability analysis of steady-state seepage by modeling spatial variability of the soil permeability [14].

Stochastic finite element methods are applied to the dynamic reliability of the structure. Successive Over Relaxation (SOR) method is used to solve the dynamic reliability of the structure. This paper presents that NSFEM computes dynamic reliability.

DYNAMIC RELIABILITY OF STRUCTURE USING THE SUCCESSIVE OVER RELAXATION METHOD

Simulation of stochastic field:

Many physics parameters of material have spatial variabilities, such as Young's modulus and Poisson's ratio, so we should regard them as stochastic fields. The stochastic field has been developed by many authors. In this paper, Young's modulus is assumed to be a Gaussian process. Improved midpoint method of stochastic field is adopted [5].

The covariance of Young's modulus between two elements that each has m nodes is obtained by:

$$Cov\left(a_{e}, a_{e'}\right) = Cov\left(\frac{a_{em1} + a_{em2} + \cdots + a_{emm} + a_{eml}}{m+1},\right.$$

$$\left.\frac{a_{e'm1} + a_{e'm2} + \cdots + a_{e'mm} + a_{e'ml}}{m+1}\right)$$

$$= \frac{1}{(m+1)^2}\left(\sum_{g_1=1}^{m}\sum_{g_2=1}^{m}Cov\left(a_{emg_1},a_{e'mg_2}\right)\right)$$

$$+\frac{1}{(m+1)^2}\left(\sum_{g_1=1}^{m}Cov\left(a_{emg_1},a_{e'ml}\right)\right)$$

$$+\frac{1}{(m+1)^2}\left(\sum_{g_2=1}^{m}Cov\left(a_{eml},a_{e'mg_2}\right)+Cov\left(a_{eml},a_{e'ml}\right)\right) \tag{1}$$

Different samples of vector $\bar{a}$ can be acquired easily (It can be seen in Chapter 4).

Dynamic reliability of structure using the Successive Over Relaxation method.

For a linear system, the dynamic equilibrium equation is given by:

$$[M]\{\ddot{\delta}\}+[C]\{\dot{\delta}\}+[K]\{\delta\}=\{F\} \tag{2}$$

where $\{\ddot{\delta}\},\{\dot{\delta}\},\{\delta\}$ are the acceleration, velocity and displacement vectors. $[M],[K]$ and $[C]$ are the global mass,

Using Wilson θ method, we obtain

$$[\tilde{K}]\{\delta_{t+\theta\Delta t}\}=\{\tilde{F}_{t+\theta\Delta t}\} \tag{3}$$

N_1 samples of vectors $a_1,a_2,\cdots,\ a_{i_1},\cdots,a_{n_1}$ are produced. N_1 matrices $\left[\tilde{K}\right]$ and N_1 Eq. 3 are generated. For linear vibrations, Eq. 3 is a system of linear equations. Successive Over Relaxation method is adopted to solve Eq. 3. It is given by:

$$\{\delta_{t+\theta\Delta t}\}^{(0)}=\left(\delta^{(0)}_{t+\theta\Delta t_1},\cdots,\delta^{(0)}_{t+\theta\Delta t_n}\right)^{T} \tag{4}$$

$$\delta^{(k+1)}_{t+\theta\Delta t_i}=\delta^{(k)}_{t+\theta\Delta t_i}+\omega\left(\tilde{F}_{t+\theta\Delta t_i}-\sum_{j=1}^{i-1}\tilde{K}_{ij}\delta^{(k+1)}_{t+\theta\Delta t_j}-\sum_{j=i}^{n}\tilde{K}_{ij}\delta^{(k)}_{t+\theta\Delta t_j}\right)\bigg/\tilde{K}_{ii} \tag{5}$$

$$(i=1,2,\cdots,n;k=0,1,\cdots)$$

It may be terminated if $\max\limits_{1\le i\le n}\left|\delta^{(k+1)}_{t+\theta\Delta t_i} - \delta^{(k)}_{t+\theta\Delta t_i}\right| < \varepsilon$ is confirmed.

where ε is the allowable error

At time $t + \Delta t$, the velocity vector and acceleration vector are obtained as:

$$\left\{\ddot{\delta}_{t+\Delta t}\right\} = b_4\left(\left\{\delta_{t+\theta\Delta t}\right\} - \left\{\delta_t\right\}\right) + b_5\left\{\dot{\delta}_t\right\} + b_6\left\{\ddot{\delta}_t\right\} \tag{6}$$

$$\left\{\dot{\delta}_{t+\Delta t}\right\} = \left\{\dot{\delta}_{t+\Delta t}\right\} + b_7\left(\left\{\ddot{\delta}_{t+\Delta t}\right\} + \left\{\ddot{\delta}_t\right\}\right) \tag{7}$$

$$\left\{\delta_{t+\Delta t}\right\} = \left\{\delta_t\right\} + \Delta t\left\{\dot{\delta}_t\right\} + b_8\left(\left\{\ddot{\delta}_{t+\Delta t}\right\} + 2\left\{\ddot{\delta}_t\right\}\right) \tag{8}$$

Vectors $\left\{\delta_{t+i_1\Delta t}\right\}, \left\{\dot{\delta}_{t+i_1\Delta t}\right\}, \left\{\ddot{\delta}_{t+i_1\Delta t}\right\}$ are solved at time $t + i_1\Delta t\left(i_1 = 2,3,\cdots,n_3\right)$ step-by -
step.

The strain and stress vectors for element d are

$$\{\varepsilon\} = [B]\{u^d\} \tag{9}$$

and

$$\{\sigma\} = [D]\{\varepsilon\} \tag{10}$$

where, $[D]$ = the material response matrix of element d, $[B]$ = the gradient matrix of
element d and $\{u^d\}$ = the element d nodal displacement vector .

The stress for element d is given by:

$$\{\sigma\} = [D][B]\{u^d\} \tag{11}$$

Substituting N_1 samples of random variables $a_1, a_2, \cdots, a_i, \cdots, a_{n_1}$ into Eq. 11, the vectors $\{\sigma\}_1, \{\sigma\}_2, \cdots, \{\sigma\}_{N_1}$ can be obtained.

The Monte Carlo simulation can compute structural reliability. A set of numbers is drawn from random variables, and the SOR is used to obtain stress value. It compares with a strength value drawn from random strength variables. If the stress is greater than the strength, the structure fails. It is supposed that simulation time is N and the failure number is F. The reliability is R=1-F/N. It is common knowledge that the Monte Carlo simulation approaches the accurate solution gradually with the increase of the number of simulations.

DYNAMIC RELIABILITY OF STRUCTURE USING NEUMANN STOCHASTIC FINITE ELEMENT

For a linear system, differential equations are transformed into linear equations by the Wilson θ method.

The stiffness matrix $\left[\tilde{\mathrm{K}}\right]$ is decomposed two matrices:

$$\left[\tilde{\mathbf{K}}\right] = \left[\tilde{\mathbf{K}}_0\right] + \left[\Delta\tilde{\mathbf{K}}\right] \tag{12}$$

where $\left[\tilde{\mathrm{K}}_0\right]$ is the stiffness matrix is replaced by mean values, $\left[\Delta\tilde{\mathrm{K}}\right]$ represents deviatoric parts. The solution $\left\{\delta_{t+\theta\Delta t}\right\}_0$ can be obtained as:

$$\left\{\delta_{t+\theta\Delta t}\right\}_0 = \left[\tilde{\mathrm{K}}_0\right]^{-1} \mathrm{F} \tag{13}$$

The Neumann expansion of $\left[\tilde{\mathrm{K}}\right]^{-1}$ takes the following form:

$$\left[\tilde{\mathbf{K}}\right]^{-1} = \left(\left[\tilde{\mathbf{K}}_0\right] + \Delta\left[\tilde{\mathbf{K}}\right]\right)^{-1} = (\mathbf{I} - \mathbf{P} + \mathbf{P}^2 - \mathbf{P}^3 + \cdots)\left[\tilde{\mathbf{K}}_0\right]^{-1} \tag{14}$$

$\left\{\delta_{t+\theta\Delta t}\right\}$ is represented by the following series as:

$$\left\{\delta_{t+\theta\Delta t}\right\} = \left\{\delta_{t+\theta\Delta t}\right\}_0 - \left\{\delta_{t+\theta\Delta t}\right\}_1 + \left\{\delta_{t+\theta\Delta t}\right\}_2 - \left\{\delta_{t+\theta\Delta t}\right\}_3 + \cdots \qquad (15)$$

This series solution is equivalent to the following equation:

$$\left[\tilde{\mathbf{K}}_0\right]\left\{\delta_{t+\theta\Delta t}\right\}_{\mathbf{i}} = \left[\Delta\tilde{\mathbf{K}}\right]\left\{\delta_{t+\theta\Delta t}\right\}_{\mathbf{i}-1}, \quad \mathbf{i} = 1, 2, \cdots \qquad (16)$$

Vectors $\left\{\delta_{t+i_1\Delta t}\right\}, \left\{\dot{\delta}_{t+i_1\Delta t}\right\}, \left\{\ddot{\delta}_{t+i_1\Delta t}\right\}$ are solved at time $t + i_1\Delta t \left(i_1 = 2, 3, \cdots, n_3\right)$ step-by–step can be seen in section 2.2.

vectors $\left\{\sigma\right\}_1, \left\{\sigma\right\}_2, \cdots, \left\{\sigma\right\}_{N_1}$ can be obtained can be seen in section 2.2.

The Monte Carlo simulation, based on NSFEM, can compute the dynamic reliability of the structure. It is supposed that the simulation time is N and the failure number is F. The reliability is R=1-F/N.

Numerical example:

(Fig. **1**) shows a cantilever beam. The load subjected to the cantilever beam is **100**sin(100t)N. It is divided into 100 eight-node quadrilateral isoperimetric elements. Its material is concrete. Young's modulus is regarded as a Gaussian process. For numerical calculation, the means of Young's modulus at each node and the midpoint within an element are $c_1\left(1.0 + \theta_1 x_i'/L\right)$. Horizontal coordinates of each node and the midpoint within an element are x_i'. The covariance of Young's modulus between any two nodes, between two midpoints and between each node and each midpoint is $c_2\left(1.0 + \theta_2 x_i/l\right)$. $c_1, c_2, \theta_1, \theta_2, l, L$ are constants. The distances between any two nodes, between two midpoints and between each node and each midpoint are x_i [5]. Table **1** shows comparisons of the NSFEM ,the SOR and the direct Monte Carlo simulation (DMCS). The NSFEM, the SOR and the DMCS separately simulate 1000 times. The DMCS uses the Cholesky decomposition to solve linear equations and provides the reference solution. The DMCS requires more CPU times than the NSFEM. The NSFEM requires more CPU times than the SOR.

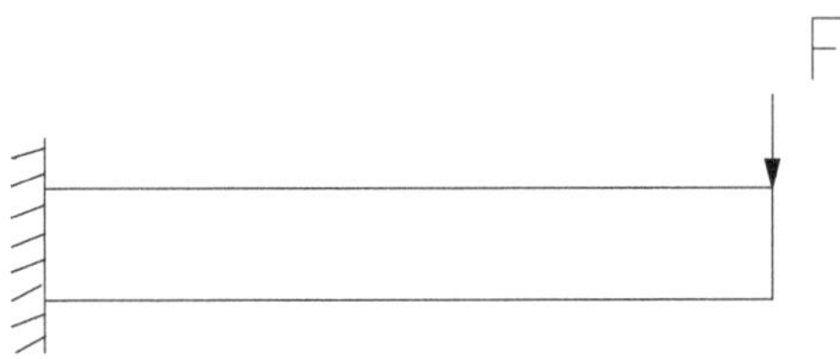

Fig. (1). A cantilever beam.

Table 1. Comparison of the NSFEM, the SOR and the DMCS.

	Reliability of Strength (Node 550)	**CPU**
NSFEM	87.1%	23 minutes 27 seconds
SOR	86.2%	20 minutes 38 seconds
DMCS	85.5%	47 minutes 6 seconds

CONCLUDING REMARKS

Considering the influence of the stochastic field, the dynamic reliability of the structure is analyzed. The Monte Carlo simulation of calculating dynamic reliability using the Successive Over Relaxation method is an efficient method. The Monte Carlo simulation of calculating dynamic reliability using the NSFEM is proposed.

CONSENT FOR PUBLICATION

Not applicable.

CONFLICT OF INTEREST

The author declares no conflict of interest, financial or otherwise.

ACKNOWLEDGEMENTS

Declared none

REFERENCES

[1]   E. Vanmarcke, *Random fields.* MIT Press: Cambridge, 1983.

[2]   W.K. Liu, T. Belytschko, and A. Mani, "Probabilistic finite elements for nonlinear structural dynamics", In: *Compu Meth Appl Mech Engrg.,* vol. 57. January, 1986, pp. 61-81.

[3]   F. Yamazaki, M. Shinozuka, and G. Dasgupta, "Neumann expansion for stochastic finite element analysis", *ASCE J. Engng. Mech.,* vol. 114, pp. 1335-1354, 1988.
[http://dx.doi.org/10.1061/(ASCE)0733-9399(1988)114:8(1335)]

[4]   S. Chakraborty, and S.S. Dey, "A stochastic finite element dynamic analysis of structures with uncertain parameters", *J Mech Sci,* vol. 40, pp. 1071-1087, 1998.
[http://dx.doi.org/10.1016/S0020-7403(98)00006-X]

[5]   W. Mo, "Stochastic finite element for structural vibration", *Math. Probl. Eng.,* vol. 2010, pp. 22-40, 2010.

[6]   S. Mahadevan, and S. Mehta, "Dynamic reliability of large frames", *Comput. Struc.,* vol. 47, pp. 57-67, 1993.
[http://dx.doi.org/10.1016/0045-7949(93)90278-L]

[7]   X.Q. Peng, L. Geng, L. Wu, G.R. Liu, and K.Y. Lam, "A stochastic finite element method for fatigue reliability analysis of gear teeth subjected to bending", *Comput. Mech.,* vol. 21.
[http://dx.doi.org/10.1007/s004660050300]

[8]   B. Sudret, and A. Der Kiureghian, "Comparison of finite element reliability methods", *Probab. Eng. Mech.,* vol. 17, pp. 337-348, 2002.
[http://dx.doi.org/10.1016/S0266-8920(02)00031-0]

[9]   H.J. Pradlwarter, and G.l. Schueller, "Uncertain linear structural systems in dynamics: Efficient stochastic reliability assessment", *Comput. Struc.,* vol. 88, pp. 74-86, 2010.
[http://dx.doi.org/10.1016/j.compstruc.2009.06.010]

[10]   W. Mo, "Monte Carlo simulation of reliability for gear", *International Conference on Computational Materials Science,* Advanced materials research, 2011pp. 42-45.
[http://dx.doi.org/10.4028/www.scientific.net/AMR.268-270.42]

[11]   M. Kamiński, and P. Świta, "Structural stability and reliability of the underground steel tanks with the Stochastic Finite Element Method", *Arch. Civ. Mech. Eng.,* vol. 15, pp. 593-602, 2015.
[http://dx.doi.org/10.1016/j.acme.2014.04.010]

[12]   H.A. Jensena, F. Mayorgaa, and C. Papadimitrioub, "Reliability sensitivity analysis of stochastic finite element models", *Comput. Methods Appl. Mech. Engrg,* vol. 296, pp. 327-351, 2015.
[http://dx.doi.org/10.1016/j.cma.2015.08.007]

[13]   X-Y. Zhou, P.D. Gosling, Z. Ullah, L. Kaczmarczyk, and C.J. Pearce, "Stochastic multi-scale finite element based reliability analysis for laminated composite structures", *Appl. Math. Model.,* vol. 45, pp. 457-473, 2017.
[http://dx.doi.org/10.1016/j.apm.2016.12.005]

[14]   A.Johari, and A. Heydari, "Reliability analysis of seepage using an applicable procedure based on stochastic scaled boundary finite element method", *Eng. Anal. Bound. Elem.,* vol. 94, pp. 44-59, 2018.
[http://dx.doi.org/10.1016/j.enganabound.2018.05.015]

CHAPTER 6

Optimization Design of Stochastic Finite Element

Abstract: The optimal design of structural vibration based on the stochastic finite element is presented. The optimization design of the gearbox is proposed by using the stochastic finite element method for structural vibration. A new iterative method (NIM) is used to calculate the stochastic finite element for vibration.

Keywords: A new iterative method, Gearbox, Optimal design, Vibration.

INTRODUCTION

Random geometric parameters, random material parameters, and random loads often exist in engineering structures. In some cases, these uncertainties deserve attention. In some cases, response to these uncertainties is so sensitive that they cannot be ignored.

Applications of random uncertainty models in the active control and structural optimization are studied [1]. Simulated annealing, clustering methods and adaptive partitioning algorithms are developed for global optimization [2]. Automation and optimization of the design of a turbine blade fir-tree root are considered [3]. The reliability-based analysis and design optimization of electrolytically-actuated Micro–Electro Mechanical Systems (MEMS) are studied [4]. Genetic algorithms (GAs) are presented to handle the fixture layout optimization problem [5]. An approach for optimization problems of non-linear FE systems under stochastic loading is presented [6]. A new optimization model based on genetic algorithms is also developed [7]. Engineering optimization of the truss-type structures using the generalized perturbation-based Stochastic Finite Element Method is proposed [8]. A discrete stochastic optimization algorithm to solve the problem of uncertain aeroelastic optimization is adapted [9]. A novel data-driven optimization strategy can efficiently handle system-level first excursion performance constraints posed on large-scale uncertain structures [10]. A universal modeling framework that encompasses all of these competing approaches is also structured [11]. A multi-objective simulation optimization algorithm that contains two crucial elements: the search phase implements stochastic kriging and the accuracy phase uses a well-known multi-objective ranking and selection procedure in view of maximizing the probability is proposed [12].

Wenhui Mo

The optimal design model of the gearbox is established by using the stochastic finite element method. A stochastic finite element is needed for the calculation of constraint conditions. A new method of stochastic finite element for vibration is proposed.

OPTIMIZATION MODEL

(Fig. **1**) shows a gearbox with 12 gears and 4 axles.

Design variables are:

$$x=(m_1,\ z_1,\ z_2,\ m_2,\ z_3,\ z_4,\ m_3,\ z_5,\ z_6,\ m_4,\ z_7,\ z_8,\ m_5,\ z_9,\ z_{10},\ m_6,\ z_{11}$$
$$,\ z_{12},\ b_1,\ b_2,\ b_3,\ b_4,\ d_1,\ l_1,\ d_2,\ l_2,\ d_3,\ l_3,\ d_4,\ l_4)^\mathrm{T}$$

The objective function is the sum of the masses of all gears and axles in the gearbox. It is defined as:

$$f(x)=\frac{1}{4}\pi b_1(m_1^2 z_1^2 - d_0^2)\rho + \frac{1}{4}\pi b_1(m_1^2 z_2^2 - d_1^2)\rho$$

$$+\frac{1}{4}\pi b_2(m_2^2 z_3^2 - d_1^2)\rho + \frac{1}{4}\pi b_2(m_2^2 z_4^2 - d_2^2)\rho$$

$$+\frac{1}{4}\pi b_2(m_3^2 z_5^2 - d_1^2)\rho + \frac{1}{4}\pi b_2(m_3^2 z_6^2 - d_2^2)\rho$$

$$+\frac{1}{4}\pi b_2(m_4^2 z_7^2 - d_1^2)\rho + \frac{1}{4}\pi b_2(m_4^2 z_8^2 - d_2^2)\rho \tag{1}$$

$$+\frac{1}{4}\pi b_3(m_5^2 z_9^2 - d_2^2)\rho + \frac{1}{4}\pi b_3(m_5^2 z_{10}^2 - d_3^2)\rho$$

$$+\frac{1}{4}\pi b_4(m_6^2 z_{11}^2 - d_3^2)\rho + \frac{1}{4}\pi b_4(m_6^2 z_{12}^2 - d_4^2)\rho$$

$$+\frac{1}{4}\pi d_1^2 l_1\rho + \frac{1}{4}\pi d_2^2 l_2\rho + \frac{1}{4}\pi d_3^2 l_3\rho + \frac{1}{4}\pi d_4^2 l_4\rho$$

where, ρ is the material density.

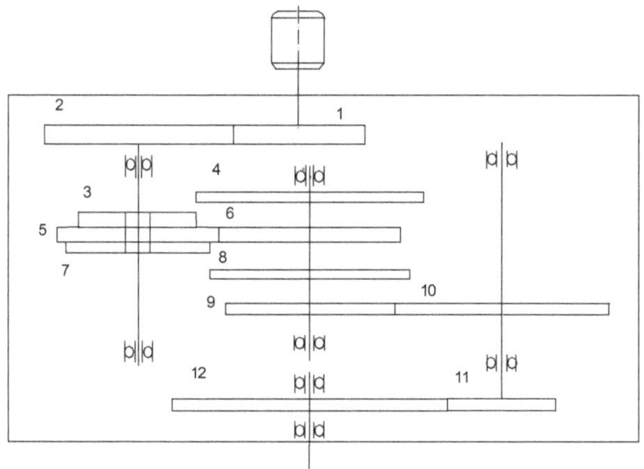

Fig. (1). A gearbox.

The constraints are defined as:

$$\overline{\sigma}_{F_i} \leq \left[\overline{\sigma}_{F_i}\right] \qquad (i=1,2,\cdots,12) \tag{2}$$

$$Var\sigma_{F_i} \leq \left[Var\sigma_{F_i}\right] \tag{3}$$

where, $\overline{\sigma}_{F_i}, Var\sigma_{F_i}$ are the mean and variance of the dynamic stress of gear bending fatigue. $\left[\overline{\sigma}_{F_i}\right], \left[Var\sigma_{F_i}\right]$ are the allowable mean value and allowable variance of gear bending fatigue strength.

$$\overline{\sigma}_{A_k} \leq \left[\overline{\sigma}_{A_k}\right] \qquad (k=I,II,III,IV) \tag{4}$$

$$Var\sigma_{A_k} \leq \left[Var\sigma_{A_k}\right] \tag{5}$$

where, $\overline{\sigma}_{A_k}, Var\sigma_{A_k}$ are the mean and variance of the dynamic stress of the dangerous section of the shaft, $\left[\overline{\sigma}_{A_k}\right], \left[Var\sigma_{A_k}\right]$ are the allowable mean value and allowable variance of shaft strength.

$$m_{kl} \leq m_k \leq m_{ks} \qquad (k=1,\ 2,\ \ldots,\ 6) \tag{6}$$

$$z_{kl} \leq z_k \leq z_{ks} \qquad (k=1,\ 2,\ \ldots,\ 12) \tag{7}$$

$$b_{kl} \leq b_k \leq b_{ks} \qquad (k=1,\ 2,\ 3,\ 4) \tag{8}$$

$$d_{kl} \leq d_k \leq d_{ks} \qquad (k=1,\ 2,\ 3,\ 4) \tag{9}$$

$$l_{kl} \leq l_k \leq l_{ks} \qquad (k=1,\ 2,\ 3,\ 4) \tag{10}$$

where, $m_{kl}, z_{kl}, b_{kl}, d_{kl}, l_{kl}$ are the lower bounds. $m_{ks}, z_{ks}, b_{ks}, d_{ks}, l_{ks}$ are the upper bounds.

Calculation of Mean and Variance of Dynamic Stress

In order to calculate the mean and variance of the dynamic stress of gear bending fatigue and the mean and variance of the dynamic stress of the shaft dangerous section, the dynamic analysis of Taylor stochastic finite element (TSFEM) is used (see Chapter 4, Ref. 7 for details). Twenty node hexahedron elements are used for the mesh of gear, and the axisymmetric quadrilateral ring element is used for the mesh of the shaft. The integrated element stiffness matrix is the whole element stiffness matrix to solve linear equations. The penalty function method is used to transform constrained optimization into unconstrained optimization, and the Powell method is used to solve unconstrained optimization. According to the optimization mathematical model of the gearbox, the computer operation program is compiled, and finally, the computer operation program is run to obtain the optimal solution.

The following methods can also be used to calculate the dynamic mean and variance of the bending fatigue stress of the gear and the dynamic stress of the dangerous section of the shaft.

Eq. 2 (chapter 4) is rewritten as:

$$Ax=b, \tag{11}$$

In this section, a new iterative method (NIM) and its convergence analysis for finding solution of (11) along with the estimation of error bounds are described. Let

$A \in Cm \times n$ and T be a subspace of Cn. Starting with $Z0 = \beta Y$, where β is a non zero real scalar, $Y \in C^{n \times m}$ satisfying $R(Y) \subseteq T$ and for any $x0 \in \_T$, the iterative method [13]is defined for $k=0,1,2,...$ by:

$$Z_{k+1} = Z_k(2I - AZ_k) \tag{12}$$

$$x_{k+1} = x_k + Z_{k+1}(b - Ax_k) \tag{13}$$

Vectors $\{\delta_{t+\Delta t}\}_1, \{\delta_{t+\Delta t}\}_2, \cdots, \{\delta_{t+\Delta t}\}_{N_1}$ are solutions of N_1 Eqs.11.

The mean of $\{\delta_{t+\Delta t}\}$ is given by:

$$\mu\{\delta_{t+\Delta t}\} = \frac{\{\delta_{t+\Delta t}\}_1 + \{\delta_{t+\Delta t}\}_2 + \cdots + \{\delta_{t+\Delta t}\}_{N_1}}{N_1} \tag{14}$$

The variance of $\{\delta_{t+\Delta t}\}$ is given by:

$$Var\{\delta_{t+\Delta t}\} = \frac{1}{N_1 - 1} \sum_{i=1}^{N_1} \left(\{\delta_{t+\Delta t}\}_i - \mu\{\delta_{t+\Delta t}\}\right)^2 \tag{15}$$

Similarly, the mean and variance of the vector $\{\delta_{t+i_1\Delta t}\}$ can be solved at time $t + i_1\Delta t \left(i_1 = 2,3,\cdots,n_3\right)$ step-by-step.

At time $t' = t + i_2\Delta t \left(i_2 = 1,2,\cdots,n_4\right)$, the strain and stress vectors for the element d are

$$\{\varepsilon\} = [B]\{\delta_{t'}^d\} \tag{16}$$

and

$$\{\sigma\} = [D]\{\varepsilon\} \tag{17}$$

where, $[D]=$ the material response matrix of the element d, $[B]=$ the gradient matrix of element d and $\{\delta_{t'}^{d}\}=$ the element d nodal displacement vector at time t'.

Substituting Eq. 16 into Eq. 17, the stress for the element d is given by:

$$\{\sigma\}=[D][B]\{\delta_{t'}^{d}\} \tag{18}$$

Substituting N_1 samples of random variables $a_1,a_2,\cdots,a_i,\cdots,a_{n_1}$ into Eq. 18, the vectors $\{\sigma\}_1,\{\sigma\}_2,\cdots,\{\sigma\}_{N_1}$ can be obtained.

The mean of $\{\sigma\}$ is given by:

$$\mu\{\sigma\}=\frac{\{\sigma\}_1+\{\sigma\}_2+\cdots+\{\sigma\}_{N_1}}{N_1} \tag{19}$$

The variance of $\{\sigma\}$ is given by:

$$Var\{\sigma\}=\frac{1}{N_1-1}\sum_{i=1}^{N_1}\left(\{\sigma\}_i-\mu\{\sigma\}\right)^2 \tag{20}$$

Comparison of Design Parameters

Using the penalty function method, a constrained optimization problem is transformed into an unconstrained optimization problem. The unconstrained optimization problem adopts the Powell method. Table **1** shows the comparison of design parameters. The TSFEM refers to the optimization parameters obtained by using the TSFEM to calculate constraints and running computer programs. The NIM refers to the optimization parameters obtained by using the NIM to calculate constraints and running computer programs. The original parameters refer to the original design parameters of the gearbox. Optimization parameters refer to the parameters of the gearbox obtained through the optimization Design of Stochastic Finite Element. After the optimization design, the gearbox is lighter in weight, smaller in volume, more reasonable, and more practical.

Table 1. The comparison of design parameters.

	m_1	m_2	m_3	m_4	z_1
Original parameters	4	4	4	4	18
Optimal parameters	3	3.5	3.5	3.5	20
	z_2	z_3	z_4	z_5	z_6
Original parameters	44	27	43	35	35
Optimal parameters	37	19	36	25	30
	z_7	z_8	z_9	z_{10}	z_{11}
Original parameters	31	39	25	41	19
Optimal parameters	23	32	19	40	19
Optimal parameters	40	40	45	46	60
	m_5	m_6	b_1	b_2	b_3
Original parameters	4	4	25	25	25
Optimal parameters	4	4	18	20	20
	b_4	l_1	l_2	l_3	l_4
Original parameters	25	350	280	340	290
Optimal parameters	25	280	190	260	190

CONCLUDING REMARKS

Considering the influence of stochastic factors, a new approach to solve the optimization design of the gearbox is presented. Using the stochastic finite element method, the optimization design of the gearbox is very efficient. The optimization design based on the stochastic finite element is applicable to mechanical products, automobiles, buildings, *etc*. The product designed by the optimization of the stochastic finite element method is light in weight, small in volume, and more competitive in the market.

CONSENT FOR PUBLICATION

Not applicable.

CONFLICT OF INTEREST

The author declares no conflict of interest, financial or otherwise.

ACKNOWLEDGEMENTS

Declared none.

REFERENCES

[1]   C. Mariani, and P. Venini, "On the use of stochastic models of uncertainty in active control and structural optimization", *Comput. Struc.,* vol. 67, pp. 105-117, 1998.
[http://dx.doi.org/10.1016/S0045-7949(97)00162-4]

[2]   L. Özdamar, and M. Demirhan, "Experiments with new stochastic global optimization search techniques", *Comput. Oper. Res.,* vol. 27, pp. 841-865, 2000.
[http://dx.doi.org/10.1016/S0305-0548(99)00054-4]

[3]   W. Song, A. Keane, J. Rees, A. Bhaskar, and S. Bagnall, "Turbine blade fir-tree root design optimization using intelligent CAD and finite element analysis", *Comput. Struc.,* vol. 80, pp. 1853-1867, 2002.
[http://dx.doi.org/10.1016/S0045-7949(02)00225-0]

[4]   M. Allen, M. Raulli, K. Maute, and D.M. Frangopol, "Reliability-based analysis and design optimization of electrostatically actuated MEMS", *Comput. Struc.,* vol. 82, pp. 1007-1020, 2004.
[http://dx.doi.org/10.1016/j.compstruc.2004.03.009]

[5]   N. Kaya, "Machining fixture locating and clamping position optimization using genetic algorithms", *Comput. Ind.,* vol. 57, pp. 112-120, 2006.
[http://dx.doi.org/10.1016/j.compind.2005.05.001]

[6]   H.A. Jensen, and A. Marillanca, "A computational procedure for response statistics-based optimization of stochastic non-linear FE-models", *Comput. Methods Appl. Mech. Eng.,* vol. 198, pp. 125-137, 2008.
[http://dx.doi.org/10.1016/j.cma.2008.02.034]

[7]   G. Corriveau, R. Guilbault, and A. Tahan, "Genetic algorithms and finite element coupling for mechanical optimization", *Adv. Eng. Softw.,* vol. 41, pp. 422-426, 2010.
[http://dx.doi.org/10.1016/j.advengsoft.2009.03.008]

[8]   M. Kamiński, and M. Solecka, "Optimization of the truss-type structures using the generalized perturbation-based Stochastic Finite Element Method", *Finite Elem. Anal. Des.,* vol. 63, pp. 69-79, 2013.
[http://dx.doi.org/10.1016/j.finel.2012.08.002]

[9]   Arnaud Rémi, "Stochastic annealing optimization of uncertain aeroelastic system", *Aerospace Science and Technology.,* vol. 36, pp. 456-464, 2014.

[10]   S. Arthriya, "Seymour, M.J.Spence, "Optimization of uncertain structures subject to stochastic wind loads under system-level first excursion constraints: A data-driven approach", *Comput. Struc.,* vol. 210, pp. 58-68, 2018.
        [http://dx.doi.org/10.1016/j.compstruc.2018.09.001]

[11]   W.B. Powell, "A unified framework for stochastic optimization", *Eur. J. Oper. Res.,* vol. 275, pp. 795-821, 2019.
        [http://dx.doi.org/10.1016/j.ejor.2018.07.014]

[12]   S.R. Gonzalez, H. Jalali, and I. Van Nieuwenhuyse, "A multiobjective stochastic simulation optimization algorithm", *Eur. J. Oper. Res.,* vol. 284, pp. 212-226, 2019.
        [http://dx.doi.org/10.1016/j.ejor.2019.12.014]

[13]   S. Srivastava, and K. Gupta Dharmendra, ""An iterative method for solving general restricted linear equations", *Appl. Math. Comput.,* vol. 262, pp. 344-353, 2015.
        [http://dx.doi.org/10.1016/j.amc.2015.04.047]

CHAPTER 7

Stochastic Finite Element for Nonlinear Vibration

Abstract: When material property geometry parameters and applied loads of structure are assumed to be stochastic, four methods of nonlinear vibration are proposed. The Newton method is applied to nonlinear vibration. Taylor expansion is used to analyze nonlinear vibration. The perturbation technology is proposed to compute nonlinear vibration. A modified iteration formula by the homotopy perturbation method is presented to analyze nonlinear vibration.

Keywords: A modified iteration formula, Nonlinear vibration, The Newton method, Taylor expansion, The perturbation technology.

INSTRUCTION

There are a lot of nonlinear vibrations in engineering problems. The influence of random factors such as random load, random geometry size, and random material characteristics on nonlinear vibration cannot be ignored. Stochastic finite element method is a good method to analyze nonlinear vibration with random parameters.

The main ingredients for developing a general-purpose version of the Spectral Stochastic Finite Element Method are proposed [1]. The application of the second-order perturbation second probabilistic moment method to the stress-based finite element method (FEM) is proposed [2]. The use of ad hoc response surface functions for the finite element analysis of uncertain structures undergoing large displacements is presented [3]. Systems modeled by elliptic partial differential equations of linear and nonlinear with stochastic coefficients (random fields) are considered [4]. Generalized nth order stochastic perturbation technique that can be applied to solve engineering with random coefficients is presented [5]. An efficient approach to the nonlinear dynamic response of structures with uncertain system properties is introduced [6]. The stability analysis of elastic systems with random parameters using the Generalized Stochastic Finite Element Method is proposed [7]. A mathematical model and its numerical realization for the composite material with stochastic interface defects are employed in the perturbation-based stochastic finite elements [8]. Integrating a model reduction technique into a simulation-based method of a class of medium/large nonlinear finite element models is presented [9].

Considering the parameters as random variables, the generalized polynomial chaos (GPC) expansion is used to analyze composite plate structures [10]. The nonlinear finite element approach is used to solve the problems of hyper thin shell roofs [11]. A macro-mechanical finite element (FE) model was developed for future performance assessments of AM structures [12]. A nonlinear proxy finite element analysis (PFEA) technique was developed, which enables the efficient prediction of bridge response up to failure [13]. A novel element-based modification of the discrete empirical interpolation has been developed [14].

The nonlinear vibration is studied by the stochastic finite element method. The Taylor expansion method, the perturbation technology, the Newton method and the modified iteration formulas by the homotopy perturbation method are proposed.

TAYLOR EXPANSION METHOD

For a nonlinear system, the dynamic equilibrium equation is given by:

$$[M]\{\ddot{\delta}\}+[C]\{\dot{\delta}\}+[K]\{\delta\}=\{F\} \tag{1}$$

where $\{\ddot{\delta}\},\{\dot{\delta}\},\{\delta\}$ are the acceleration, velocity and displacement vectors. $[M],[K]$ and $[C]$ are the global mass, stiffness and damping matrices obtained by assembling the element variables in the global coordinate system.

By using the Newmark method, Eq. 1 becomes

$$\{\delta_{t+\Delta t}\}=\left[\tilde{K}\left(\delta_{t+\Delta t}\right)\right]^{-1}\left\{\tilde{F}_{t+\Delta t}\right\} \tag{2}$$

where, $\{\delta_{t+\Delta t}\}$, $\left[\tilde{K}\right]$ and $\left\{\tilde{F}_{t+\Delta t}\right\}$ indicate the displacement vector, stiffness matrix and load vector at time $t+\Delta t$.

Eq. 2 is rewritten as:

$$\left[\tilde{K}\left(\delta_{t+\Delta t}\right)\right]\{\delta_{t+\Delta t}\}=\left\{\tilde{F}_{t+\Delta t}\right\} \tag{3}$$

Many physics parameters of material possess spatial variabilities, such as Young's modulus and Poisson's ratio, so we should regard them as stochastic fields. Means of Young's modulus and Poisson's ratio represent Eq. 3. The Newton method is used to solve nonlinear equations so that δ is obtained. After δ represents Eq. 3, it becomes a linear equation containing stochastic variables of Young's modulus and Poisson's ratio.

Eq. 3 is rewritten as:

$$\left[\tilde{K}\right]\left\{\delta_{t+\Delta t}\right\} = \left\{\tilde{F}_{t+\Delta t}\right\} \tag{4}$$

The partial derivative of Eq. 4 with respect to a_i is given by:

$$\frac{\partial\left\{\delta_{t+\Delta t}\right\}}{\partial a_i} = \left[\tilde{K}\right]^{-1}\left(\frac{\partial\left\{\tilde{F}_{t+\Delta t}\right\}}{\partial a_i} - \frac{\partial\left[\tilde{K}\right]}{\partial a_i}\left\{\delta_{t+\Delta t}\right\}\right) \tag{5}$$

where,

$$\frac{\partial\left\{\tilde{F}_{t+\Delta t}\right\}}{\partial a_i}$$ can be found in chapter 4.

After $\dfrac{\partial\left\{\delta_t\right\}}{\partial a_i} = q_0, \dfrac{\partial\left\{\dot{\delta}_t\right\}}{\partial a_i} = \dot{q}_0, \dfrac{\partial\left\{\ddot{\delta}_t\right\}}{\partial a_i} = \ddot{q}_0$ are given, Eq. 5 can be calculated.

The partial derivative of Eq. 5 with respect to a_j is given by

$$\frac{\partial^2\left\{\delta_{t+\Delta t}\right\}}{\partial a_i \partial a_j} = \left[\tilde{K}\right]^{-1}\left(\frac{\partial^2\left\{\tilde{F}_{t+\Delta t}\right\}}{\partial a_i \partial a_j} - \frac{\partial\left[\tilde{K}\right]}{\partial a_i}\frac{\partial\left\{\delta_{t+\Delta t}\right\}}{\partial a_j}\right.$$

$$\left. - \frac{\partial\left[\tilde{K}\right]}{\partial a_j}\frac{\partial\left\{\delta_{t+\Delta t}\right\}}{\partial a_i} - \frac{\partial^2\left[\tilde{K}\right]}{\partial a_i \partial a_j}\left\{\delta_{t+\Delta t}\right\}\right) \tag{6}$$

where,

$$\frac{\partial^2 \left\{ \tilde{F}_{t+\Delta t} \right\}}{\partial a_i \partial a_j}$$ can be found in chapter 4 .

After $\dfrac{\partial \{\delta_t\}}{\partial a_j} = q_1,\ \dfrac{\partial \{\dot{\delta}_t\}}{\partial a_j} = \dot{q}_1,\ \dfrac{\partial \{\ddot{\delta}_t\}}{\partial a_j} = \ddot{q}_1,\ \dfrac{\partial^2 \{\delta_t\}}{\partial a_i \partial a_j} = r_0,\ \dfrac{\partial^2 \{\dot{\delta}_t\}}{\partial a_i \partial a_j} = \dot{r}_0\ ,\ \dfrac{\partial^2 \{\ddot{\delta}_t\}}{\partial a_i \partial a_j} = \ddot{r}_0$ are given, Eq. 6 can be calculated.

The displacement is expanded at the mean value point $\overline{a} = \left(\overline{a}_1, \overline{a}_2, \cdots, \overline{a}_i, \cdots, \overline{a}_{n_1} \right)^T$ by means of a Taylor series. The mean of $\delta_{t+\Delta t}$ is obtained as:

$$\mu\{\delta_{t+\Delta t}\} \approx \{\delta_{t+\Delta t}\}\big|_{a=\overline{a}} + \frac{1}{2}\sum_{i=1}^{N}\sum_{j=1}^{N} \frac{\partial^2 \{\delta_{t+\Delta t}\}}{\partial a_i \partial a_j}\bigg|_{a=\overline{a}} Cov(a_i, a_j) \tag{7}$$

where, $\mu\{\delta_{t+\Delta t}\}$ expresses mean value $\delta_{t+\Delta t}$ and $Cov(a_i, a_j)$ is the covariance between a_i and a_j.

The variance of $\delta_{t+\Delta t}$ is given by:

$$Var\{\delta_{t+\Delta t}\} \approx \sum_{i=1}^{N}\sum_{j=1}^{N} \frac{\partial \{\delta_{t+\Delta t}\}}{\partial a_i}\bigg|_{a=\overline{a}} \cdot$$

$$\frac{\partial \{\delta_{t+\Delta t}\}}{\partial a_j}\bigg|_{a=\overline{a}} \cdot Cov(a_i, a_j) \tag{8}$$

$$\frac{\partial \{\ddot{\delta}_{t+\Delta t}\}}{\partial a_i}\ ,\ \frac{\partial \{\dot{\delta}_{t+\Delta t}\}}{\partial a_i}\ ,\ \frac{\partial^2 \{\ddot{\delta}_{t+\Delta t}\}}{\partial a_i \partial a_j}$$ and $$\frac{\partial^2 \{\dot{\delta}_{t+\Delta t}\}}{\partial a_i \partial a_j}$$ can be found in chapter 4 .

$$\frac{\partial\{\ddot{\delta}_{t+\Delta t}\}}{\partial a_i}, \frac{\partial\{\dot{\delta}_{t+\Delta t}\}}{\partial a_i}, \frac{\partial^2\{\ddot{\delta}_{t+\Delta t}\}}{\partial a_i \partial a_j} \text{ and } \frac{\partial^2\{\dot{\delta}_{t+\Delta t}\}}{\partial a_i \partial a_j} \text{ must be calculated for the following}$$

iteration.

Then, the mean and variance of displacement are obtained at time $t + i_1\Delta t\left(i_1 = 2, 3, \cdots, n_1\right)$ step by step.

Perturbation Technology

See the above section for the mathematical treatment of the transformation of nonlinear equations into linear equations.

$\left[\tilde{K}\right], \{\delta_{t+\Delta t}\}$ and $\{\tilde{F}_{t+\Delta t}\}$ are expanded about a_i *via* Taylor series.

$$\left[\tilde{K}\right] = \left[\tilde{K}\right]^0 + \sum_{i=1}^{n}\frac{\partial\left[\tilde{K}\right]}{\partial a_i}\alpha_i + \sum_{i=1}^{n}\frac{\partial^2\left[\tilde{K}\right]}{\partial a_i^2}\alpha_i^2 + \cdots \tag{9}$$

$$\{\delta_{t+\Delta t}\} = \{\delta_{t+\Delta t}\}^0 + \sum_{i=1}^{n}\frac{\partial\{\delta_{t+\Delta t}\}}{\partial a_i}\alpha_i + \sum_{i=1}^{n}\frac{\partial^2\{\delta_{t+\Delta t}\}}{\partial a_i^2}\alpha_i^2 + \cdots \tag{10}$$

$$\{F_{t+\Delta t}\} = \{F_{t+\Delta t}\}^0 + \sum_{i=1}^{n}\frac{\partial\{F_{t+\Delta t}\}}{\partial a_i}\alpha_i + \sum_{i=1}^{n}\frac{\partial^2\{F_{t+\Delta t}\}}{\partial a_i^2}\alpha_i^2 + \cdots \tag{11}$$

By applying the perturbation technique, the following equations are given by:

$$\{\delta_{t+\Delta t}\}^0 = \left(\left[\tilde{K}\right]^0\right)^{-1}\{F_{t+\Delta t}\}^0 \tag{12}$$

$$\frac{\partial\{\delta_{t+\Delta t}\}}{\partial a_i} = \left[\tilde{K}\right]^{-1}\left(\frac{\partial\{\tilde{F}_{t+\Delta t}\}}{\partial a_i} - \frac{\partial\left[\tilde{K}\right]}{\partial a_i}\{\delta_{t+\Delta t}\}\right) \tag{13}$$

$$\frac{\partial^2 \{\delta_{t+\Delta t}\}}{\partial a_i \partial a_j} = \left[\tilde{K}\right]^{-1} \left(\frac{\partial^2 \{\tilde{F}_{t+\Delta t}\}}{\partial a_i \partial a_j} - \frac{\partial \left[\tilde{K}\right]}{\partial a_i} \frac{\partial \{\delta_{t+\Delta t}\}}{\partial a_j} \right.$$

$$\left. - \frac{\partial \left[\tilde{K}\right]}{\partial a_j} \frac{\partial \{\delta_{t+\Delta t}\}}{\partial a_i} - \frac{\partial^2 \left[\tilde{K}\right]}{\partial a_i \partial a_j} \{\delta_{t+\Delta t}\} \right) \tag{14}$$

where $\dfrac{\tilde{\partial} \{F_{t+\Delta t}\}}{\partial a_i}$ and $\dfrac{\partial^2 \{\tilde{F}_{t+\Delta t}\}}{\partial a_i^2}$ can be found in chapter 4.

The mean of the displacement is obtained as:

$$\mu_\delta \approx \{\delta_{t+\Delta t}\}\Big|_{a=\bar{a}} + \frac{1}{2} \sum_{i=1}^{n} \frac{\partial^2 \{\delta_{t+\Delta t}\}}{\partial a_i^2}\Big|_{a=\bar{a}} \cdot \sigma_i^2 \tag{15}$$

The variance of $\{\delta_{t+\Delta t}\}$ is given by:

$$Var_\delta \approx \sum_{i=1}^{n} \left(\frac{\partial \{\delta_{t+\Delta t}\}}{\partial a_i}\Big|_{a=\bar{a}} \right)^2 \cdot \sigma_i^2 \tag{16}$$

$\dfrac{\partial \{\ddot{\delta}_{t+\Delta t}\}}{\partial a_i}$, $\dfrac{\partial \{\dot{\delta}_{t+\Delta t}\}}{\partial a_i}$, $\dfrac{\partial^2 \{\ddot{\delta}_{t+\Delta t}\}}{\partial a_i \partial a_j}$ and $\dfrac{\partial^2 \{\dot{\delta}_{t+\Delta t}\}}{\partial a_i \partial a_j}$ can be found in chapter 4 .

The mean value and variance of the displacement are obtained at time $t + i_1 \Delta t \left(i_1 = 2, 3, \cdots, n_3 \right)$ step-by-step.

Newton Method

Using the covariance matrix, the correlation of Young's modulus between any two elements is given by:

$$C_{aa} = \begin{pmatrix} Cov(a_1,a_1) & Cov(a_1,a_2) & \cdots & Cov(a_1,a_N) \\ Cov(a_2,a_1) & Cov(a_2,a_2) & \cdots & Cov(a_2,a_N) \\ \vdots & \vdots & & \vdots \\ Cov(a_N,a_1) & Cov(a_N,a_2) & \cdots & Cov(a_N,a_N) \end{pmatrix} \tag{17}$$

A Gaussian vector $\bar{a} = [a_1,a_2,\cdots,a_N]^T$ is generated

$$\bar{a} = LZ \tag{18}$$

$Z = [Z_1,Z_2,\cdots,Z_N]^T$ consists of N Gaussian random variables with mean zero and unit standard deviation. The Cholesky matrix L can be obtained through a decomposition of the covariance matrix, therefore

$$\mu\left[ZZ^T\right] = I \tag{19}$$

$$LL^T = C_{aa} \tag{20}$$

I is the identity matrix. The generation of vector $\bar{a}$ must satisfy the covariance matrix

$$\mu\left[\bar{a}\bar{a}^T\right] = \mu\left[LZ(LZ)^T\right]$$

$$= L\mu\left[ZZ^T\right]L^T = C_{aa} \tag{21}$$

Once the decomposition has been completed, different samples of vector $\bar{a}$ can be acquired easily by Eq. 18. Thus, it is possible that Monte Carlo simulation resolves the problem of stochastic finite element.

Using Eq18, N_1 samples of vector $\bar{a}$ are produced. N_1 matrices $\left[\tilde{K}\right]$ and N_1 Eqs.3 are generated. For nonlinear vibrations, Eq. 3 is a system of nonlinear equations. The Newton method is an adequate method for solving large systems of nonlinear equations. It can be accomplished as follows:

Eq. 3 is rewritten as:

$$F(x) = 0 \tag{22}$$

The solution of Eq. 22 is given by:

$$x^{(k+1)} = x^{(k)} - F'(x^{(k)})^{-1} F(x^{(k)}) \quad (\text{k=0, 1}, \cdots) \tag{23}$$

where,

$$F'(x) = \begin{pmatrix} \dfrac{\partial f_1(x)}{\partial x_1} & \dfrac{\partial f_1(x)}{\partial x_2} & \cdots & \dfrac{\partial f_1(x)}{\partial x_n} \\[2mm] \dfrac{\partial f_2(x)}{\partial x_1} & \dfrac{\partial f_2(x)}{\partial x_2} & \cdots & \dfrac{\partial f_2(x)}{\partial x_n} \\[2mm] \vdots & \vdots & & \vdots \\[2mm] \dfrac{\partial f_n(x)}{\partial x_1} & \dfrac{\partial f_n(x)}{\partial x_2} & \cdots & \dfrac{\partial f_n(x)}{\partial x_n} \end{pmatrix} \tag{24}$$

Vectors $\{\delta_{t+\Delta t}\}_1, \{\delta_{t+\Delta t}\}_2, \cdots, \{\delta_{t+\Delta t}\}_{N_1}$ are solutions of N_1 Eqs.3.

The mean of $\{\delta_{t+\Delta t}\}$ is given by:

$$\mu\{\delta_{t+\Delta t}\} = \frac{\{\delta_{t+\Delta t}\}_1 + \{\delta_{t+\Delta t}\}_2 + \cdots + \{\delta_{t+\Delta t}\}_{N_1}}{N_1} \tag{25}$$

The variance of $\{\delta_{t+\Delta t}\}$ is given by:

$$Var\{\delta_{t+\Delta t}\} = \frac{1}{N_1 - 1} \sum_{i=1}^{N_1} \left(\{\delta_{t+\Delta t}\}_i - \mu\{\delta_{t+\Delta t}\}\right)^2 \tag{26}$$

Similarly, the mean and variance of vector $\{\delta_{t+i_1\Delta t}\}$ can be solved at time $t + i_1\Delta t$ step by step.

The Homotopy Perturbation Method

Using Eq. 18, N_1 samples of vector $\bar{a}$ are produced. N_1 matrices $\left[\tilde{K}\right]$ and N_1 Eqs.3 are generated. For nonlinear vibrations, Eq. 3 is a system of nonlinear equations. A modified iteration formula by the homotopy perturbation method (MIHPD) with accelerated fourth-and fifth-order convergence [15] is used to solve Eq. 3.

Eq. 3 is rewritten as:

$$\Phi(X) = \begin{cases} f_1(X) \\ f_2(X) \\ \vdots \\ f_N(X) \\ X = (X_1, X_2, \cdots X_N)^T \in \Re^N \end{cases} \tag{27}$$

The solution of Eq. 3 is given by:

$$y_{i,m} = x_{i,m} - \sum_{n=1}^{N} \Psi_{i,n}(X_m) f_n(X_m) \tag{28}$$

$$z_{i,m} = -\sum_{n=1}^{N} \Psi_{i,n}(Y_m) f_n(Y_m) \tag{29}$$

$$x_{i,m} = y_{i,m} + z_{i,m} - \frac{1}{2} \sum_{n=1}^{N} \sum_{j=1}^{N} \sum_{k=1}^{N} \Psi_{i,n}(X_m) \frac{\partial^2 f_n(Y_m)}{\partial x_j \partial x_k} z_{j,m} z_{k,m} \tag{30}$$

$$i = 1, 2, \cdots, N, m = 0, 1, \cdots$$

Vectors $\{\delta_{t+\Delta t}\}_1, \{\delta_{t+\Delta t}\}_2, \cdots, \{\delta_{t+\Delta t}\}_{N_1}$ are solutions of N_1 Eqs.3.

The mean of $\{\delta_{t+\Delta t}\}$ is given by:

$$\mu\{\delta_{t+\Delta t}\} = \frac{\{\delta_{t+\Delta t}\}_1 + \{\delta_{t+\Delta t}\}_2 + \cdots + \{\delta_{t+\Delta t}\}_{N_1}}{N_1} \tag{31}$$

The variance of $\{\delta_{t+\Delta t}\}$ is given by:

$$Var\{\delta_{t+\Delta t}\} = \frac{1}{N_1 - 1}\sum_{i=1}^{N_1}\left(\{\delta_{t+\Delta t}\}_i - \mu\{\delta_{t+\Delta t}\}\right)^2 \tag{32}$$

Similarly, the mean and variance of vector $\{\delta_{t+i_1\Delta t}\}$ can be solved at time

$t + i_1\Delta t$ step by step.

(Fig. **1**) shows a cantilever beam. The length is 1m, the width is 0.2m, and the height is 0.05m. The load subjected to the cantilever beam is 100sin(100t) N. Its material is concrete. The cantilever beam vibrates nonlinearly. It is divided into 40 rectangle elements. Young's modulus is regarded as a stochastic process. Table **1** shows the mean and variance of vertical displacement at node 55 when the cantilever beam has vibrated for six seconds. The DMCS indicates direct Monte Carlo simulation. The DMCS simulates 1000 samples. It is common knowledge that the DSFEM approaches the accurate solution gradually with the increase in the number of simulations. The TE indicates the Taylor expansion method of nonlinear vibration analysis. The PT indicates the perturbation technology of nonlinear vibration analysis. The NM indicates the Newton method of nonlinear vibration analysis. The MIHPD indicates a modified iteration formula by the homotopy perturbation method of nonlinear vibration analysis.

Fig. (1). A cantilever beam.

Table 1. The comparisons of the mean and variance of vertical displacement at node 55.

	Mean(mm)	Variance(mm^2)
DSFMC	0.043	0.00076
TE	0.047	0.00097
PT	0.047	0.00097
NM	0.045	0.00084
MIHPD	0.044	0.00081

CONCLUDING REMARKS

Considering the influence of random factors, four stochastic finite element methods are proposed to calculate nonlinear vibration. The calculation formulas are given. An example shows that the four methods are correct. The four methods in this chapter are applicable to nonlinear vibration analysis of mechanical products, automobiles, buildings, *etc.*

CONSENT FOR PUBLICATION

Not applicable.

CONFLICT OF INTEREST

The author declares no conflict of interest, financial or otherwise.

ACKNOWLEDGEMENTS

Declared none.

REFERENCES

[1]   R. Ghanem, "Ingredients for a general purpose stochastic finite elements implementation", *Comput. Methods Appl. Mech. Eng.,* vol. 168, pp. 19-34, 1999.
[http://dx.doi.org/10.1016/S0045-7825(98)00106-6]

[2]   Marcin Kaminski, "Stochastic second-order perturbation approach to the stress-based finite element method", *International Journal of Solids and Structures,* vol. 38, pp. 3831-3852, 2001.
[http://dx.doi.org/10.1016/S0020-7683(00)00234-1]

[3]   N. Impollonia, and A. Sofi, *"A response surface approach for the static analysis of stochastic structures with geometrical nonlinearities",* Computer Methods in Applied Mechanics and Engineering., vol. 192. Desember, 2003, pp. 4109-4129.

[4]   H.G. Matthies, and A. Keese, "Galerkin methods for linear and nonlinear elliptic stochastic partial differential equations", *Comput. Methods Appl. Mech. Eng.,* vol. 194, pp. 1295-1331, 2005.
[http://dx.doi.org/10.1016/j.cma.2004.05.027]

[5]   M. Kamiński, "Generalized perturbation-based stochastic finite element method in elastostatics", *Comput. Struc.,* vol. 85, pp. 586-594, 2007.
[http://dx.doi.org/10.1016/j.compstruc.2006.08.077]

[6]   GeorgeStefanou, "MichalisFragiadakis, "Nonlinear dynamic analysis of frames with stochastic non-Gaussian material properties", *Eng. Struct.,* vol. 31, pp. 1841-1850, 2009.
[http://dx.doi.org/10.1016/j.engstruct.2009.02.020]

[7]   M.M. Kamiński, and P. Świta, "Generalized stochastic finite element method in elastic stability problems", *Comput. Struc.,* vol. 89, pp. 1241-1252, 2011.
[http://dx.doi.org/10.1016/j.compstruc.2010.08.009]

[8]   M. Kamiński, and J. Szafran, "Perturbation -based stochastic finite element analysis of the interface defects in composites *via* Response Function Method", *Compos. Struct.,* vol. 97, pp. 269-276, 2013.
[http://dx.doi.org/10.1016/j.compstruct.2012.10.023]

[9]   H.A. Jensen, F. Mayorga, and C. Papadimitriou, "Reliability sensitivity analysis of stochastic finite element models", *Computer Methods in Applied Mechanics and Engineering,* vol. 296, pp. 327-351, 2015.
[http://dx.doi.org/10.1016/j.cma.2015.08.007]

[10]   K. Sepahvand, "Stochastic finite element method for random harmonic analysis of composite plates with uncertain modal damping parameters", *J. Sound Vibrat.,* vol. 400, pp. 1-12, 2017.
[http://dx.doi.org/10.1016/j.jsv.2017.04.025]

[11]   Arghya Ghosh, "Dipankar Chakravorty, "First ply failure analysis of laminated composite thin hypar shells using nonlinear finite element approach", *Thin-walled Struct.,* vol. 131, pp. 736-745, 2018.
[http://dx.doi.org/10.1016/j.tws.2018.07.046]

[12]   F. Parisi, C. Balestrieri, and H. Varum, "Nonlinear finite element model for traditional adobe masonry", *Construction,* vol. 223, pp. 450-462, 2019.
[http://dx.doi.org/10.1016/j.conbuildmat.2019.07.001]

[13]   A.P. Schanck, and W.G. Davids, "Capacity assessment of older t-beam bridges by nonlinear proxy finite-element analysis", *Structures,* vol. 23, pp. 267-278, 2020.
[http://dx.doi.org/10.1016/j.istruc.2019.09.012]

SUBJECT INDEX

T

V

www.ingramcontent.com/pod-product-compliance
Lightning Source LLC
Chambersburg PA
CBHW042041110726
48006CB00002B/251